ACHIEVING BUSINESS AGILITY

STRATEGIES FOR BECOMING PIVOT READY IN A DIGITAL WORLD

John Orvos

Achieving Business Agility: Strategies for Becoming Pivot Ready in a Digital World

John Orvos
Belle Mead, New Jersey, USA

ISBN-13 (pbk): 978-1-4842-3854-7
ISBN-13 (electronic): 978-1-4842-3855-4
https://doi.org/10.1007/978-1-4842-3855-4

Library of Congress Control Number: 2018955308

Managing Director, Apress Media LLC: Welmoed Spahr
Acquisitions Editor: Susan McDermott
Development Editor: Laura Berendson
Coordinating Editor: Rita Fernando

Distributed to the book trade worldwide by Springer Science+Business Media New York, 233 Spring Street, 6th Floor, New York, NY 10013. Phone 1-800-SPRINGER, fax (201) 348-4505, e-mail orders-ny@springer-sbm.com, or visit www.springeronline.com. Apress Media, LLC is a California LLC and the sole member (owner) is Springer Science + Business Media Finance Inc (SSBM Finance Inc). SSBM Finance Inc is a **Delaware** corporation.

For information on translations, please e-mail rights@apress.com, or visit http://www.apress.com/rights-permissions.

Apress titles may be purchased in bulk for academic, corporate, or promotional use. eBook versions and licenses are also available for most titles. For more information, reference our Print and eBook Bulk Sales web page at http://www.apress.com/bulk-sales.

Any source code or other supplementary material referenced by the author in this book is available to readers on GitHub via the book's product page, located at www.apress.com/9781484238547. For more detailed information, please visit http://www.apress.com/source-code.

Printed on acid-free paper

To my wife and kids, who were patient while I stared and clicked away on my laptop writing this during evenings and weekends over the past two years.

Contents

About the Author

John Orvos brings techniques from over 20 years' experience as a business consultant, agile practice leader, and award-winning agile strategist. John has led transformations alongside hundreds of agile consultants dealing with adopting and scaling agile in Fortune 500 organizations. Over the past two years, John has been challenged by these customers to accomplish even more ambitious goals—an agile business—in order for them to better compete against digital disruptor competitors. John provides a fresh business-oriented perspective on how to enable an entire organization to achieve business agility.

About the Technical Contributors

Deema Dajani led some of the largest transformations to business, portfolio, and delivery agility in Fortune 500 companies. She helped several financial services companies transform to adaptive organizations that better sense and respond to market disruption. She started the agile journey in the early 2000s and never looked back, at a fast-growth startup that was one of the earliest success stories for adopting scrum across the enterprise. Drawing on her MBA education from the Kellogg School, Deema advises on go-to-market strategy. She carries her passion into the local community, cofounding agile groups and the Women in Agile nonprofit.

Marie Kalliney is a transformational, results–oriented business and information technology leader with 20 years of progressive experience, focused on global software portfolio, program, and product management. She drives business and IT transformations and bridges both communities to optimize processes and deliver value from technology investments. Marie has a strong passion for the business and commitment to the development of solutions to complex business issues to enable successful organizations.

Yvonne Kish Delaney is a Sr. Principal Agile Consultant at CA Technologies (formerly Rally Software). The agile movement in the early 2000s was a game changer, and Yvonne realized that early on and never turned back. Since then agile has been at the foundation of Yvonne's coaching in enabling companies to get truly excited about engaging and working in new ways to efficiently deliver high-quality products that are of great value to their customers. Yvonne specializes in large, complex organizational agile transformations. Leveraging her agile subject matter expertise coupled with deep software engineering and quality background, Yvonne has led numerous companies in translating their product vision into market-ready reality.

Gene Mrozinski has been in the business of planning, designing, writing, testing, and deploying software for over four decades. During that time, he has seen a lot of change, but none so fundamental and important as agile. It has changed the game in so many ways. He is fortunate enough to have been involved in agile for almost 20 years now and enjoys working with teams and companies in not only adopting agile but helping them continue their journey into what comes next. Areas like scaling, agile portfolio management, and true business agility are some of the focus areas that he is passionate about.

Ronica L. Roth is a practice development lead for agile services at CA Technologies (formerly Rally) who coaches the coaches. She helps design and bring to market the services that will help client companies become better workplaces and corporate citizens. Since the early 2000s, she has been trying to push the boundaries of "lean" and "agile" to touch more people and places. She facilitates groups of people to have great interactions and outcomes, and she coaches executives to adopt an agile mindset as they lead culture change.

Preface

There was a loud buzz in the room. It was standing room only during the sold-out New York City business agility summit in the spring of 2017, where over 400 executives and thought leaders packed into the conference center. I had high expectations to learn something new about the emerging "agile business" frontier and was happy to pay for the $1200 ticket. The highly anticipated keynote speaker, a well-known agile thought leader, set the tone by asserting that business agility success depends on "having the right culture and mindset of teamwork" within an organization, and that "business executive support" is key to agile business success. I got that; I silently agreed, since I've heard these agile principles many times.

But then, as the day went on, a string of agile transformation speakers took turns merely echoing those same basic ideas throughout the day. Wait. That's it? I thought. Not one speaker was attempting to define business agility or explain strategies to accomplish this? To the contrary, I was hearing the same worn-out success stories about various departments implementing agile ways of working with the words "agile business" cut and pasted in their presentation slides. What made it more astonishing was all the Kool-Aid drinking and groupthink at the event. It seemed like no one else felt the same, and my efforts to share this view kept falling on deaf ears. I felt like I was in a time warp and back in 2001. Has nothing evolved since then?

A great deal of attention has been given to the idea of business agility, but comparatively little insight has been offered into what it really means or strategies that create the right team collaborative culture and business executive buy-in. Consequently, even after over fifteen years of agile evangelism, consulting, and adoption efforts, these agile ways of working remain largely within

IT, R&D, GIS, software development, and product delivery (which we will call "delivery"[1]), while organizations struggle to expand and adopt them into the business. As a result, delivery can claim departmental success with agile ways of working to improve software development but is still falling short of the true potential to enhance the organization overall. Agile, as we know it, is suspended in mediocrity; it has not lived up to its true promise.

In a CA Technologies survey of 150 executives between March and June 2017, only 12% of participants felt their entire organization is on the path to business agility ("The State of Business Agility," 2017).[2] The majority (58%) of respondents cited cultural or political barriers as primary challenges in fully adopting agile methodology.

This book will help guide an organization's adoption of business agility. In these pages, I will introduce clear, actionable steps for modernizing and utilizing change management truths that I believe have been long overlooked, thereby limiting the success of agile expansion into the business. My aim is not to pile on reasons for why organizations need to expand agile into the business or to spout recycled notions about having a collaborative and team culture. Those subjects are well covered by oceans of articles, books, and events. Rather, this

[1]Delivery—short for "product delivery"—comprises all the groups who contribute to building any software or solution that interacts with customers, including the those who build the enabling technologies that ultimately contribute to that goal. Delivery might include:

- a product delivery or product management group, which studies customers and the market and defines the vision and roadmaps for the products and solutions we sell (including technology products, technology-enabled products, and nontech products)
- a product management office, which manages the company's product and/or technology investments
- a research and delivery (R&D) or engineering or technology department, which builds the software that comprises or supports the solutions
- all or part of an information technologies (IT) or general information systems (GIS) department, which builds the enabling technologies that support all of these activities

A note on the internal IT or GIS department. Some of their work does not contribute to building product but rather supports other business processes, like around selling, marketing, finance, and so forth. Although that work is not part of product delivery, the company still benefits from being more pivot ready if IT is agile.

[2]Gatepoint Research, "Business Agility Throughout the Enterprise Pulse Report," May 2017, www.ca.com/content/dam/ca/us/files/ebook/the-state-of-business-agility-2017.pdf.

book will provide a pragmatic framework for driving the business toward an agile mindset and facilitating cultural changes that will help your organization deliver great products and services.

Achieving Business Agility offers four strategies and concrete examples for engaging the business executives and will teach you how to execute these strategies effectively. Whether you are a delivery executive, change advocate, consultant, business leader, or newcomer to agile, you will learn clear actions from a pragmatic, business-oriented perspective that is vital to effecting change and bringing agility into the business.

Acknowledgments

This has been a team effort.

Coaches Call-Out Contributors

Saif Islam, Agile Managing Consultant, CA Technologies

Christen McLemore, Agile Managing Consultant, CA Technologies

Chris Browne, Senior Director, Agility Services, CA Technologies

Rob Desmarais, Senior Director, Agility Services, CA Technologies

Editorial Contributors

Gary Putlock, Senior Director, Agile Management, CA Technologies

John Martin, Director, Agile Services, CA Technologies

Alfred Aversa, Agile Solution Director, CA Technologies

Joe Katusha, Agile Solution Director, CA Technologies

Jim Rezen, Agile Solution Director, CA Technologies

Brian Piening, Agile Solution Director, CA Technologies

Monica Russ McCarthy, Agile Solution Director, CA Technologies

Chris Pola, Principal Agilist, CA Technologies

Jason Deno, Agile Advisor, CA Technologies

Suzanne Lorch, Agile Strategist, CA Technologies

Rob Kowalski, Agile Solution Director, CA Technologies

Brent Chalker, Global Account Director, CA Technologies

Greg Kiey, Agile Solution Director, CA Technologies

Wendy Schall, Agile Manager, CA Technologies

Jordan Goodwin, Agile Manager, CA Technologies

Introduction

Whether you are a technology or business executive, you certainly know that agile development has been practiced in your company's delivery department for some time. Over the past 15 years, organizations across the globe have made major investments to implement, scale, and expand these practices within delivery. You probably also recognize the importance of extending this agile way of working beyond delivery and into the business to fend off the emergence of digital disruptor competitors: doing so is vital for your company to compete in today's marketplace. By connecting agile ways of working between delivery and the business, your company can better understand its customers and quickly pivot on its product go-to-market strategy to serve their fast-changing needs. A fully agile business can sense market changes and quickly respond with competitive product offerings. As a result, it will compete more effectively because it can change direction on a dime to morph its product offerings and capitalize on market opportunities or thwart potential threats. The definition of business agility to be used in this book, therefore, is the ability to sense and respond as a matter of everyday business (or, as the subtitle of this book says, becoming "pivot ready"). With the increasing numbers of digital disruptors, the ability of a company to sense market opportunities and threats and respond quickly with products that customers value is vital.

Sense and respond has been one of the original selling points for delivery to adopting agile ways of working. Yet does anyone feel it has lived up to its potential? Are companies that adopt agile ways of working really able to sense and respond and deliver customer value faster than the competition? Regrettably, the answer is almost always no. Agile ways of working have not yet achieved their fullest potential in organizations. That's because agile ways of working are trapped inside delivery. Although this helps delivery to deploy products fast, the company is not aligned to support it. So, delivery builds products fast based on guesses of what customers want and, in turn, pushes out products that have no go-to-market plan from the business. Agile, as we know it, is stifled in delivery and falling short from realizing its full, untapped potential to help organizations sense and respond to compete more effectively.

Figure I-1 shows the evolution of agile through the years.

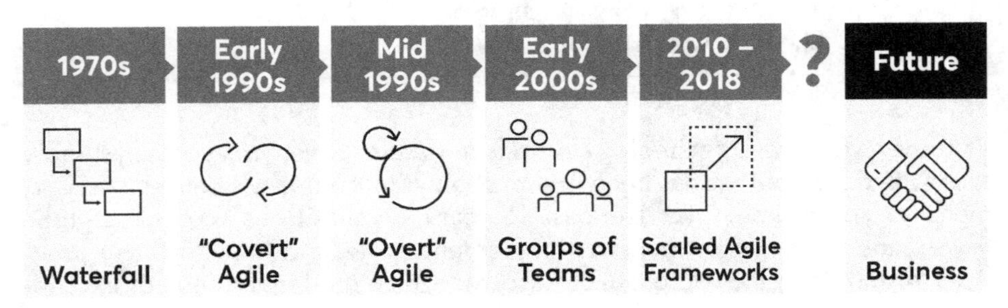

Figure I-1. Evolution of agile

Waterfall software development era: 1970s–80s

- A noniterative and sequential approach, as progress flows in largely one direction ("downward" like a *waterfall*) through the phases of conception, initiation, analysis, design, construction, testing, deployment, and maintenance.

Covert agile software development era: early 1990s

- A new approach for software *development* under which requirements and solutions evolve through the collaborative effort of self-organizing cross-functional teams.
- Rapid application development (James Martin), adaptive software development (Jim Highsmith, Sam Bayer), lean principles (James Womak and Daniel Jones) Scrum (Ken Schwaber, Jeff Sutherland), XP (Ken Beck, Ward Cunningham, Ron Jeffries), and others.

Overt agile software (team-level) development era: mid-1990s

- Came out of frustration in the 1990s. The enormous time lag between business requirements (the applications and features customers were requesting) and the delivery of technology that answered those needs led to project overruns or the canceling of many projects. Business requirements and customer requisites changed during this lag time, and the final product did not meet the then-current needs. The software development models of the day, led by the Waterfall model, were not meeting the demand for speed and did not take advantage of just how quickly software could be altered.

Groups of agile team development era (no Framework): early 2000s

- Agile Manifesto (2001) is a milestone that brought together seventeen agile thought leaders to proclaim the value in the following principles: individuals and interactions over processes and tools; working software over comprehensive documentation; customer collaboration over contract negotiation; responding to change over following a plan. Team-level iterative agile started to gain momentum.
- Kanban, a manufacturing method from the 1940s, was first embraced for software development also at the team level. Instead of iterative, Kanban is focused more on continuous flow mainly by limiting work in progress.

Scaled agile framework development era: 2010 to today

- As agile ways of working have continued to expand in popularity, there has been a growing need for agile frameworks that will work for large-scale programs and projects in the enterprise. Frameworks promote alignment, collaboration, and delivery across large numbers of agile teams. The primary reference for the scaled agile framework was originally the development of a big-picture view of how work flowed from product management, through governance, program teams, and development teams, out to customers.
- Large multiteam scaling methods started to emerge, with frameworks such as Scaled Agile Framework (SAFe), Large-Scale Scrum (LeSS), Disciplined Agile Delivery (DAD), and Nexus

Future

- Scaling to portfolio agility, not just program levels
- The start of business agility,which is transforming whole businesses to be adaptive to change

DEFINING BUSINESS AGILITY

Is business agility the same as agile "within" the business?

In the various agile conferences with a gathering of all the agile thought leaders and consultants, it has become clear that there is confusion about what business agility means. Many of the so-called success stories are being told based on practicing agile in a particular business unit, such as HR, marketing, compliance, and so on. Their claim of success centers around practicing agile ceremonies within these departments to better run their workflow. As a result, they improve their productivity, morale, teamwork, and transparency to better manage their work within their respective departments. That's great for the department, but it's not business agility. Instead, this is the definition of implementing agile within the business.

Achieving business agility is far different than implementing agile within the business. An agile business connects everything back to developing software products that customers value. So, all activities are in sync with the common goal to develop and deploy the most valuable product. Business agility is focused on improving the company's ability to sense and respond to changes in the marketplace. Therefore, business activities are connected to building products that provide value to the customer. An emerging mistake is getting the terms confused with agile within the business. Agile within the business refers to when there is the adoption of agile practices in the business without any connection to building software products.

Although a noble cause, agile within the business is not the same because it has nothing to do with helping the company compete in the marketplace. An agile business has the single mission to help the company sense and respond to change in order to compete by delivering high-value software products. Therefore, business agility is different from agile within the business.

So where to begin? We could launch right into a long list of agile success stories about improved productivity, quality, and meeting software delivery deadlines. But let's acknowledge something: achieving business agility goes far beyond building software in delivery using agile practices, so venturing into uncharted territory outside the delivery comfort zone may feel like crossing into a different world.

Imagine you are an astronaut who has successfully orchestrated many trips to the moon—agile in delivery. Now, in the coming months, you must prepare for a mission to Mars—agile in the business. While you have years of experience to build on your moon endeavors, the rules will be different for your upcoming journey to Mars. The conditions, coordinates, equipment, and even language you've been using may not work for this new planet. When a fellow astronaut approaches you and asks, "So, how are we going to get to Mars?," the last thing you would want to do is smugly reply, "Since we are experts at going to the moon, we will simply take off in that direction and hope we get lucky and find the right way."

To reach your new destination, there's no question that you will need a new plan. Unfortunately, many agile experts, who have had enormous success in delivery, assume that their current model can simply be dusted off and reused for new missions. Though I have attended many agile classes over the past ten years, I don't recall a single class, speech, or discussion about change management strategy as it applies to business adoption. In fact, I have often heard that the success of agile delivery assumes that the business side of the organization would magically migrate to adopt these same practices and come to the table; explaining nothing about strategies to communicate what's in it for the business to make this change. Change management has never been on the radar screens of the agile transformation consulting world because change management teams' well-oiled routines have been working smoothly in delivery over the years—they know how to get to the moon. Now with the new challenge to reach Mars, how can we expect success if we don't know how to get there? Why do we talk so little about business change management strategies instead of dusting off the old ones that have worked in delivery? To start this journey, we must first acknowledge that the business is different than delivery and, therefore, requires a business-oriented change management approach to adopt agile ways of working. This change management approach must answer the looming question that various business executives will be thinking, which is, "What's in it for me?"

Despite the increasing importance of business agility for companies to stay competitive these days, agile software development practices are too often stuck in the delivery box, remaining disconnected from the business. A fundamental problem occurs when delivery executives set out to convince their business to join their agile crusade but are repeatedly denied due to misconceptions. There is, however, a way to bring the organization together under a common goal.

Change management suffers in the agile world because agile has been well known in delivery and generally accepted. Within the business, there is minimal understanding about agile, or even worse is that it's completely misunderstood, and that often results in high levels of resistance to applying agile beyond delivery. Therefore, an agile change management framework that resonates in a business context is vital. Unfortunately, those who have successfully led agile delivery transformations are sometimes overconfident when working with the business have the tendency of not earning the trust, interest, and buy-in of business executives. So, the agile evangelists have remained stuck in delivery, often entrenched in the mechanics of scaling agile. They fail to leverage their success in delivery when talking about agile in a business context. Without a real strategy for bringing everyone along for the necessary changes, many of today's agile transformation experts simply hope that there is some business executive in the organization that intuitively understands the value proposition and will actively endorse their agile journey. But hope is not a strategy. Hence the purpose for this book: to offer an actionable change management strategy and framework for any employee to use within their organization and is an alternative to just betting on hope.

How This Book Is Structured

This book is divided into three parts, but you can read them in any order you choose. If you want to start with understanding the strategies before seeing them in action, read the parts in sequential order. Part 1 is instructional and can be re-read as a reference during your journey toward business agility. Part 2 includes a riveting story that demonstrates how the strategies discussed in his book can be applied to a situation. Part 3 then ties everything together with final thoughts to send you on the right path to success. Feel free to bounce around and read with flexibility.

Part 1: The Mechanics - Understanding Strategies to Achieve Business Agility. Chapters 1-4 lay out the strategies to sell business agility's vision within your company and gain executive acceptance.

Part 2: The Story - Strategies in Action. Stories about applying strategies to the business. Chapters 5–10 center on the fictional case study of Linda and the Solar Corona Insurance company. I'll show you how they implemented the strategies discussed in Part 1 to advance change toward business agility.

Part 3: The Takeaway - Defining Your Company's Journey. Chapters 11 and 12 bring it all together so you can start your journey toward business agility.

PART

1

The Mechanics - Understanding Strategies to Achieve Business Agility

CHAPTER

1

Sound the Alarm

A sound the alarm strategy will get an executive's attention by alerting him or her to an imminent company problem and creates a sense of urgency to address it. This strategy is based on getting your executive's attention by pointing out the ramifications of an emerging business problem: new competitors are winning over your customers and are poised to take the lead. Today, companies are competing on a whole new playing field. Customer demands are changing too fast for most companies to keep up. As the world becomes more digital, traditional companies are being disrupted and are rapidly losing market share to more agile competitors. Modern companies need to rethink how they will adjust in order to survive. In today's competitive landscape, speed matters. If companies cannot respond quickly—with more agility—they will be left behind.

When an executive does not have the *awareness* that the company is facing new tough challenges from competitive entry and increasing customer demands, it's time to sound the alarm and create a sense of *urgency* to pay attention to this problem. The reality is that your company's executives are always busy and, in some cases, they feel as though everything is going just fine. They may be thinking, "I've been successful thus far, so I must be doing things right. I am constantly getting requests from people wanting something. Why should I spend time dealing with yet another person asking me to do something when I already have too much work to do?"

J. Orvos, *Achieving Business Agility*, https://doi.org/10.1007/978-1-4842-3855-4_1

Your Alarm: Disruptors Are Winning

Your company is sure to be facing the problem of digital disruption by faster, more agile competitors that are cranking out new products to seize opportunities created from more demanding customers. In the past, when opportunities appeared in the market, your big-fish company (see Figure 1-1) would simply exert its brute force and gobble them up. But today is different. Digital disruptors are everywhere, and they quickly respond by attacking a single aspect of or weakness in your company's product and improving upon it to create their own solution. Once this competitive solution succeeds with customers, your company has missed a key opportunity. If you can't keep up, the digital disruptors will move on to attack the next part of your company's blind spot in the market.

Your company in the past

Your company today

Figure 1-1. Big fish company

Digital disruptors are competitors who are transforming their businesses by building their solutions quickly to meet customer needs. For a company to not only survive, but also thrive, it must eliminate the constraints between these customer-facing pillars—in other words, become a disrupter rather than the one being disrupted.

Although organizations have been changing for years, the current pace of change has accelerated with disruptor entry, and organizations no longer have the luxury of procrastination or resting on their laurels. Unlike the past, digital disruptors are constantly flooding from all directions, so organizations must be built to change and able to rapidly pivot their direction to thrive in today's digital world.

Being first to market in the digital world has huge competitive ramifications, or in the case of established players or incumbents, being a "fast follower" is a matter of survival. Software innovation is everywhere, and it's vital to sense and respond fast in order to survive and win against digital disruptors.

Key Questions: Defining Your Company's Problem

When preparing to sound the alarm, your focus is to reveal the company problem, which is it's lack of ability to accurately understand or sense market changes and effectively respond to and execute a product's go-to-market strategy in a timely manner. In other words, is your business able to change quickly on it's go-to-market strategy?

When defining your company's problem, research answers to these two questions regarding competitive and customer pressures.

1. What are the new competitive pressures for your company?

   What chatter does the industry have about disrupting your space? Who are your company's biggest emerging digital disruptor competitors? How large of a threat do they pose? Can these digital disruptors sense and respond to customer demands and develop a unique solution that would pose a competitive threat to your company? What are your company's key competitive advantages and disadvantages in responding faster than these digital disruptors?

2. What are the new customer pressures for your company?

   What are today's customer needs that are no longer met by your company's solutions? What is your company's current level of customer satisfaction? Has that been improving or declining over the past year? How are customer expectations changing? What new trends are emerging in other industries that could impact your customers? How is your company sensing these changes in order to meet the expectations of more demanding customers? What patterns from your customer feedback can help you predict a future disruption, drop-off, complaints, and lost deals?

A recent study by Innosight, "Corporate Longevity: Turbulence Ahead for Large Organizations," illustrates how Fortune 500 companies are being disrupted by smaller, more agile companies.[1] The findings warn business executives that about half of the S&P 500 will be replaced over the next 10 years. Some notable companies that have already been dropped from the S&P 500 over the last six years are Kodak, Sprint, Abercrombie & Fitch, JCPenney, RadioShack, Dell, Avon, The New York Times, and Safeway.

To paint a clear picture of the pressures your company is facing, you can point to case studies based on your experience. A number of companies have run into problems that they had the chance to avoid, but they chose to stay the course and failed. Companies that were once massive leaders in their market, like Blockbuster, Kodak, and TomTom, ended up going bankrupt by ignoring the digital disruptors. If your company executives have the same naive lack of awareness, you may be on your way to a similar fate.

DISRUPTOR GRAVEYARD

1. **Blockbuster** refused to change and stuck to their retail store business model, disregarding new customer desires. They lost to Netflix.

Blockbuster dominated the video rental business. Founded in 1985, Blockbuster offered movies and games for rent and purchase. In 2004, Blockbuster appeared the unbeatable behemoth of the industry with about 9,000 retail stores globally. Although successful, a large part of Blockbuster revenue came from late fees, which they liked, but the customer started to resent.

[1]Scott. D. Anthony, S. Patrick Viguerie, Andrew. Waldeck, "Corporate Longevity: Turbulence Ahead for Large Organizations," Innosight Executive Briefing, Spring 2016, www.innosight.com/wp-content/uploads/2016/08/Corporate-Longevity-2016-Final.pdf.

The main competitor, Netflix, started as a DVD by-mail subscription service and went public in 2002. Its founder, Reed Hastings, started Netflix partly out of frustration after being fined $40 by Blockbuster for being late in returning "Apollo 13." Netflix internet and subscription services eliminated late fees and emerged to challenge Blockbuster.

Initially, Blockbuster was unaware that late fees became an important enough issue for customers to cause them to switch to Netflix mail order for movie rentals. By 2004, Netflix had already cut into Blockbuster's customer base and Blockbuster finally noticed, although too late. After they became aware, and sensed Netflix as a threat, Blockbuster had yet another barrier to overcome—their ability to respond quickly. Blockbuster simply could not respond with an answer fast enough.

Blockbuster did not have the ability to sense what customers wanted and respond with a full go-to-market offering. As a result, they were unable to answer this challenge. By 2010, Blockbuster had to file for bankruptcy protection, moving to wipe out around $1 billion in debt. By 2013, Blockbuster announced plans to close its remaining US stores.

2. **Kodak** stayed analog while their customers moved digital.

Kodak had tremendous market power share, controlling almost 70% of the US film market. It did not respect the significance of the digital movements in 1981, even though they helped to create the new digital medium that would eventually displace them.

Kodak was faced with an influx of digital disruptors, which were a completely different business in their view. Kodak sensed these competitors but was reluctant to change its mindset and reinvent itself. Simply, Kodak wasn't willing to part with the legacy that had made them successful. The company, in its comfortable position of dominance, was slow to respond and neglected to create new digital products that met the change markets demand and provide customer value.

In 2012, Kodak filed for Chapter 11 bankruptcy protection.

3. **TomTom** focused on creating software for stand-alone GPS devices rather than capitalizing on the entrance, emergence, and dominance of smartphone devices.

In 2002 TomTom was successful and growing by selling its GPS software to run on portable navigation devices. In 2007–08, the iPhone suddenly and unexpectedly appeared as competition in the market. These smartphones had their own GPS and navigation systems, directly competing with TomTom. TomTom failed to respond quickly and entered into a crisis in 2009. The stock price plummeted while sales and market share began a steady decline until the present day.

Other examples of large companies having to rethink how they compete against emerging digital competitors include American Express, Visa, and MasterCard—who worry less about competing with one another these days and more about how to defend against Apple Pay, Google Wallet, and a slew of digital disruptor startups popping up around the world. Insurance companies like Cigna, Aetna, and UnitedHealth Group understand the rules of the game for competing with one another; what they must figure out now is how to manage the onslaught of unknown digital disruptors on what used to look like a level playing field. The list goes on in every industry, with customers looking to do more Internet banking, searching for consumer products online, engaging in telecommunications, and more.

Target the Four Pillars

There are so many departments and executives who are like minded and who want to be part of the solution. Fending off the competition is absolutely critical, so do you simply seek any executive who will listen? Do you speak to executives you already know? Do you make overtures to random department heads? One agile leader from a Fortune 100 US bank said, "I asked around and found an executive who was interested in sponsoring a pilot. I got lucky."

Luck may lead to a small win at times, but it is far from an assurance of reaching your ultimate goals. Can you get to Mars by launching in that general direction and hoping you'll get lucky? Perhaps your agile business campaign will have a degree of success in improving speed but will deliver products that are not valued by the customer, or maybe you will not be able to move the full organization to deal with the company problem of digital competitors and demanding customers. For your company to compete and solve these problems, it needs to do more than just increase the speed of delivery. If it were just about speed, then you could simply stay within just the delivery department and push out software faster by automating processes in development operations (DevOps). But improving speed is not directly correlated to creating products that your target market and customers want. In this case, your delivery is creating the product fast. So, it's not just about speed to deliver, but rather speed to deliver products that customers value, or "speed to value." In turn, in order to deliver value, your delivery department must know what customers want. Otherwise, your company might increase speed of delivery, but not speed to value. You might be building the wrong things fast.

There is only one way to do this: delivery, as one pillar, needs to be connected to these other customer-facing pillars: marketing, finance, and channels. So, when preparing to engage business leaders in agility practices, you must target these pillars. Instead of hoping to get lucky, set a strategic course, beginning with a proactive and deliberate approach to these four pillars.

The Four Customer-Facing Pillars

Delivery, finance, marketing, and **channels** are the four customer-facing departments of any organization that determine the company's ability to compete in the marketplace (Figure 1-2). These departments have always been considered important to a company's competitive viability since the 1960s. They are called the four Ps—product, price, promotion, and place. These have been tried and true over the years as the pillars of any company to compete: having the best product quality (product), at market-competitive price (price), with a proper marketing messaging (promotion), and distribution channels to the customer (place). Therefore, the four customer-facing departments will be defined as the four pillars in this book.

- Delivery refers to the department that builds software product and solutions that customers can use. Delivery consists of software developers, project management, program management, and product (owner) strategy. Product strategists who also have a product manager title are a proxy for customers and make key decisions about what products to build.
- Finance, marketing, and channels are the customer-facing departments that we will define as the business. Finance funds the products to be developed by delivery, marketing promotes these products, and channels sells and supports these products; all of their decisions touch the customer directly.

The activities of the four departments directly (or indirectly for finance) impact customers and, therefore, have the greatest influence on their decisions to buy or not buy from the company. As a result, these departments have the most significant effect on the company's viability and, therefore, carry the most weight to move the needle of organizational change. They are the tip of the spear with the customer and are key to determining the competitiveness of any company. Given their importance, if they change, then they become the "winds of change" in the sails for the rest of the organization. All other contributing departments, including HR, legal, compliance, operations, training, and field support are important, but they will ultimately align with these customer-facing departments and follow their lead.

Four Pillars

Companies compete by having the best-quality **product** at a fair **price**, with enticing **promotions** and accessible **places** for customers to buy.

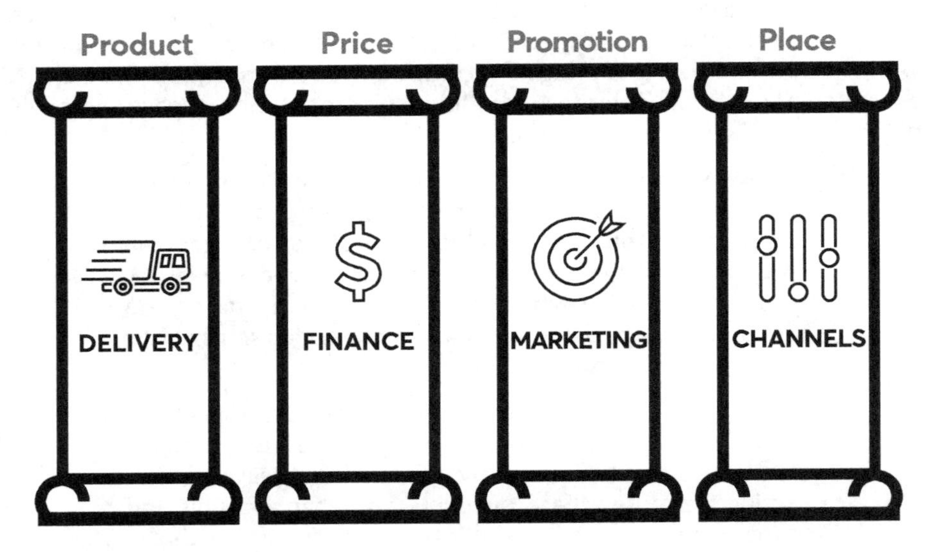

Figure 1-2. Four pillars

HOW SOFTWARE IS CHANGING THE GAME

Although these four pillars remain consistent today, they need to be modernized to deal with the new challenges businesses are facing in the digital world. More than ever, they are impacted by software in today's application economy—where customers use and consume software, and products powered by software: product (delivery builds software products), price (finance funds software that impacts pricing decisions), promotion (marketing uses software for messaging), and place (channels use software to interact with the customers). As you begin your business agility initiative, you should target executives from these pillars.

Software is now a vital part of any company's success and impacts your company's ability to deliver value to the customer and compete on the new digital playing field. For your company to stay relevant, your company must now count on software to drive the business in ways they never had to in the past. Therefore, as the person sounding the alarm, you are doing your executives the favor of helping to highlight a problem before it's too late to address it. The question is not "if" they have a problem, but rather "when and how" they will need to do something about it. Do they want to be proactive and get ahead of it or do they want to wait and react when agile digital competitors start winning?

The Change in Delivery

Traditionally, all companies aim to create unique products that fulfill customer needs and wants. Whether producing razor blades, clothing, detergent, cars, or insurance products, a company exerts tremendous effort to generate the best goods and services and to differentiate them from the offerings of other companies in the market.

Today, software (built by delivery) is being incorporated into core products and plays a major part in their viability. Though invisible to the user, embedded software is vital to product development as it enables the networking of devices, vehicles, clothing, home appliances, buildings, and more.[2]

It is fundamental to supporting the Internet of Things (IoT)[3]; therefore, the executives who lead product development are critical targets for a business agility transformation. For purposes of this report, we will define a product as any software or software-embedded product.

In order for these software products to keep up with the pace of change, new methods, techniques, and tools have been created to enable delivery groups to respond quickly by building or enhancing "just enough" of a product. These methods, techniques, and tools have created a new game for delivery that focuses on delivering smaller increments of intense value quickly, while balancing quality and the longer-term view.

---

[2]**Tesla Motors** integrates software into almost every aspect of the cars they manufacture. As a result, drivers can receive notifications when the oil needs to be changed (based on precise readings of viscosity and peak temperature) and when it is time for the annual tune-up; mechanics can be alerted when the dreaded check engine light comes on, and appointments can be proposed automatically based on cross-checks of digital calendars, easily confirmed with a click on a smartphone. What's more, software updates can occur "over the air," while the cars sit in their owners' garages or driveways or in the parking lot at work, just as smartphone upgrades happen remotely. These features need not be subscription-based; they are increasingly included in the basic experience of owning the car.
Nest depends on software in their production of Internet-connected home devices such as thermostats, smoke detectors, and cameras. Google bought Nest in 2014 for $3.2 billion.
[3]**IoT**, the Internet of Things, enables everyday objects to be operated through embedded software. "Smart" heating systems and air conditioning units, garage doors, appliances, lighting, security devices, and more can be controlled through the Internet.

The Change in Finance

Traditionally, during annual budgeting ceremonies the finance department allocates funding to the various projects based on the request from the product management.[4]

Since delivery is focused on delivering smaller increments of value more quickly, while balancing quality and the longer-term view, the game is changing for finance when it comes to investing. Certainly, the macro-level view of investing in products, customer value streams, and operating models is still very important, but the micro-level view has become equally–if not more–critical. Today, finance needs to fund the software products that are highly valued by the customer, which are moving targets. Since the customer demands and taste for software change fast, finance does not have the luxury of waiting each year in between decisions. Funding the right software innovations must keep pace with the quickly changing customer demands. By properly funding solutions that are highly valued by the customer, the business can set higher prices and not be bound to setting market competitive prices.

The Change in Marketing

Traditionally, marketing develops and communicates the proper messages about company products to potential customers. Through advertising, public relations, social media, e-mail, search engines, videos, and more, the marketing department generates product awareness, interest, and desire.

In 2018 and beyond, software is vital to promoting products and services. Customers use countless Web, social media, and mobile applications to learn about what they want to buy. Often software drives the primary form of brand communication between the customer and the organization.

But more importantly, this software-enabled marketing is changing the game by enabling customers to tell the brand story–for good or bad–to thousands and sometimes millions of other potential and existing customers. This direct link that marketing has to the customer via their phone, tablet, or PC has also enabled the collection of intelligence about that customer and their wants and needs. Marketing can no longer afford to simply target segments or groups that roughly hit the mark: they need to tailor their messages directly to each customer.

---

[4]The product's price affects how it sells since customers compare it against competitive offerings. As a result, the lines of business set pricing to be at par with the competitive products with slight increases or decreases staying in line with its perceived customer value. Therefore, finance needs to make annual bets to funds the right products that will perform well in the marketplace and deliver business results. If the product has weak results in the marketplace, than pricing may have to be set lower to encourage customers to buy on price. If the product is stronger, the price may be set higher with that added value to the customer.

The Change in Channels

Traditionally, customers purchase products or services through distribution and sales channels. Channels convert potential interest into actual customers. Enabling customers to engage with products and make purchases happens through various channels: sales associates, advisors or consultants, merchandising, and customer service or support. These channels connect with customers in many venues such as a retail store (brick and mortar), trade shows, offices, call centers, catalogs, and in the field.

Today, in addition to the traditional channels, new digital channels that drive product sales have emerged, such as e-commerce websites and mobile apps. And many companies are implementing software that incorporates feedback directly into their products, both by making it easier to provide feedback in-product and by introducing mechanisms for measuring interaction and other metrics.

Channels interact with the customer through Web and mobile applications and social media as part of our everyday lives. These digital channels are providing a seamless customer experience, whether via bricks and mortar, online or mobile, or catalog or social media. In addition, the self-service functions online represent opportunities to generate revenue as well as reduce labor costs for companies.

One cannot overlook the impact digital channels have had on traditional brick and mortar in the retail space. Once established players such as Toys "R" Us, Borders, and JCPenney have all lost their way by not adopting digital in time.

Software in channels has changed the game by enabling the mantra of "anytime, anywhere, any means" when it comes to delivering products and services to customers. In addition to enabling that access through a specific means, customers have also come to demand that–however their interaction occurs–their experience be seamless. For example, they want to be able to use their phone to order a product, call someone to discuss an issue, get updates on that issue through their phone, and–if necessary–go to a brick-and-mortar store to return or exchange a product, with the store having full access to the purchase and service history, no matter which store they visit.

Helping Pillar Executives See the Impacts

Now that you've homed in on the company problem, you need to understand how your business executives in each of the four pillars—delivery, finance, marketing, and channels—might be affected personally. Ask yourself:

- How does the company problem impact each executive in their particular leadership role?
- What losses or gains will each executive leader face based on how the company problem is handled?
- How would the failure to handle the company problem impact each executive in each pillar negatively?

You are looking for specific answers to these questions that apply to a particular executive in each pillar. These answers will be your basis for formulating a hypothesis or an impact statement about a looming problem that will affect the executive in each pillar.

Because the company problem may be commonly known across your company, it is your job to find out something the executive does not know. There are always blind spots. They may have misread the scope, complexity, or duration of the situation they are facing. For example, an executive may believe something will be accomplished quickly, when in reality, it will take much longer and be more expensive and problematic. By uncovering this and alerting the executive, you will solidify your position.

Developing an Impact Statement

Using your acquired intelligence about company pressures and executive drivers, craft an impact statement to use during your early conversations for executives in each of the four customer-facing pillars who are unaware there is a problem. Your statement will gently provoke any unknowing and unaware executive to realize that there is a potentially unanticipated problem that he or she will care about.

Remember, this impact statement is when you sound the alarm that a problem exists. And to reiterate from the introduction of this chapter, this sound-the-alarm strategy is for those executives that do not know about the company's competitive and customer problem. However, if they do have awareness about this company problem, then you do not need to sound the alarm and, instead, should select another strategy.

An impact statement highlights the company problem along with its executive drivers that connect with each executive's interest from the four customer-facing pillars, depending on their role and line of business. You will be speaking to executives from various pillars across the organization, and they all have different goals. For each pillar executive, understand their fears, such as fear of failure or missing out on an opportunity to succeed. But frame the conversation positively: you are simply offering the executive a chance to be proactive, honing their attention on the company's problem early so they can feel confident in getting ahead of it.

KEY TIP Succeed in getting attention by looking out for the business executive, not by thinking about what they can do for you.

Prepare a 20-second WE-YOU statement designed to sound the alarm. For example, to convey the message that *WE are facing a company problem*, you might say:

> *As you may know, our customers are changing (such as…), and emerging competitors (such as…) are starting to make moves to attract them (such as...). Our company now has to determine how to respond to these new challenges.*

To then convey how YOU will be impacted as an executive, you might say:

> *Based on my work in the field, research on our emerging competitors, and input from more demanding customers, I would like to share new insights about how ...*

Tailoring your impact statements to match the interests and needs of each of the respective pillars will make your conversations even more successful. Here are some examples of what you might say as you sound the alarm.

Alarm for Delivery

"Although delivery is already agile, customers are not using the software much—**competitors** have higher usage." (See Figure 1-3.)

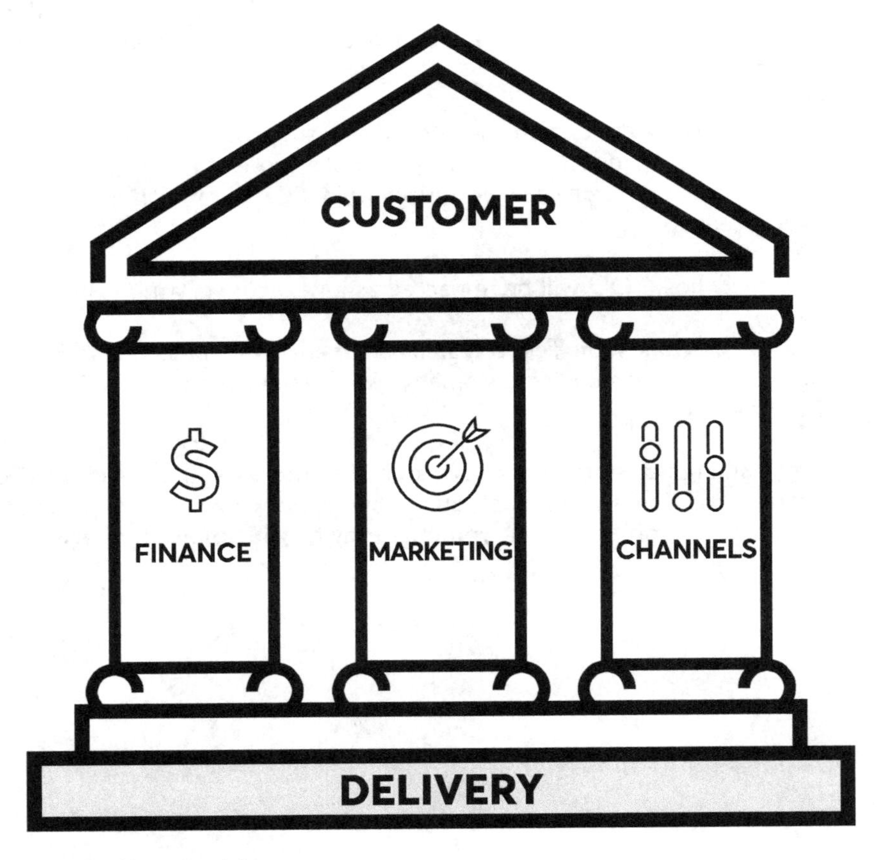

Figure 1-3. Alarm for delivery

Alarm for Finance

"**Competitors have higher** profit margins with innovations." (See Figure 1-4.)

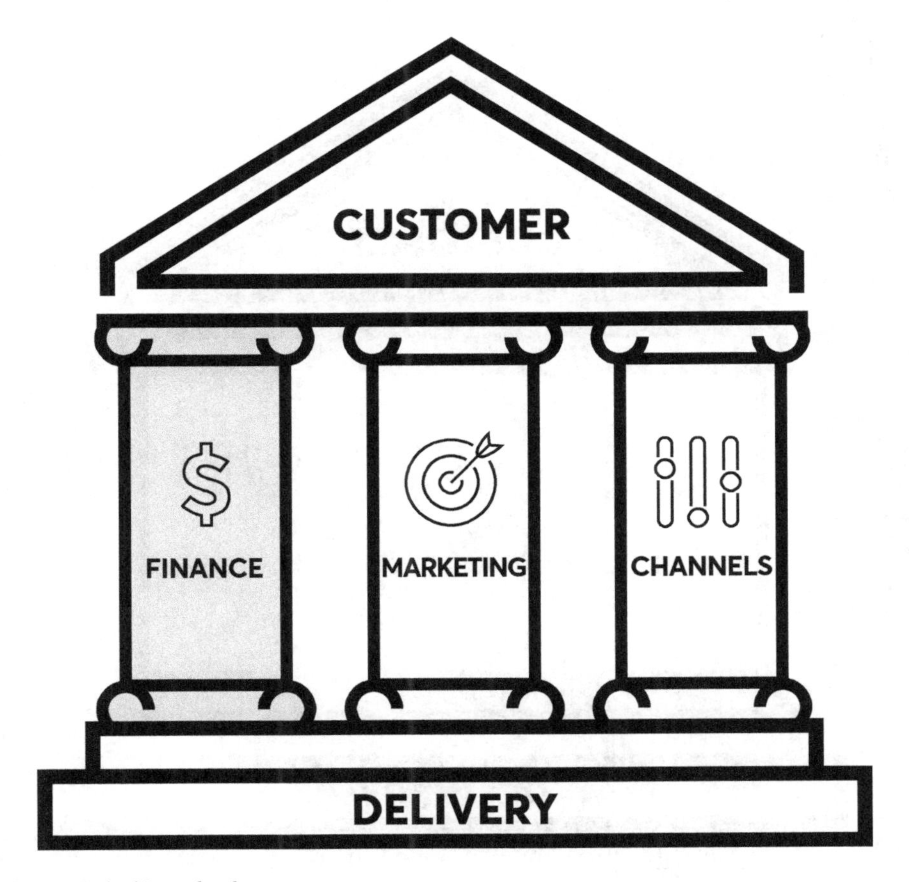

Figure 1-4. Alarm for finance

Alarm for Marketing

"**Competitors are attracting** a growing new pool of quality leads because their messaging is relevant in comparison." (See Figure 1-5.)

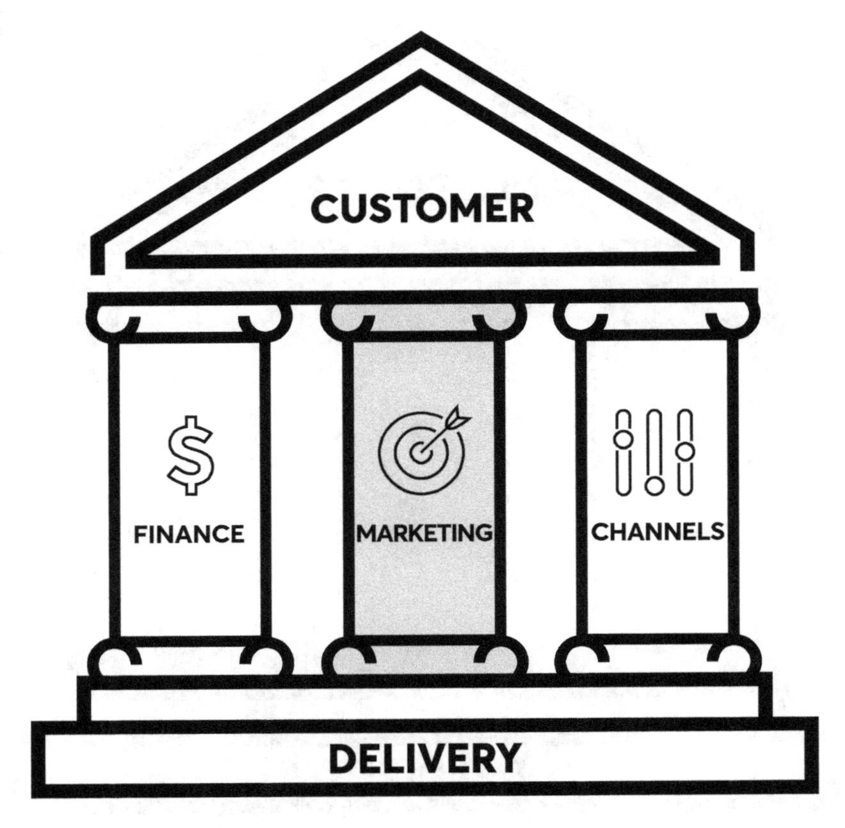

Figure 1-5. Alarm for marketing

Alarm for Channels

"**Competitors are increasing** sales significantly with new channels." (See Figure 1-6.)

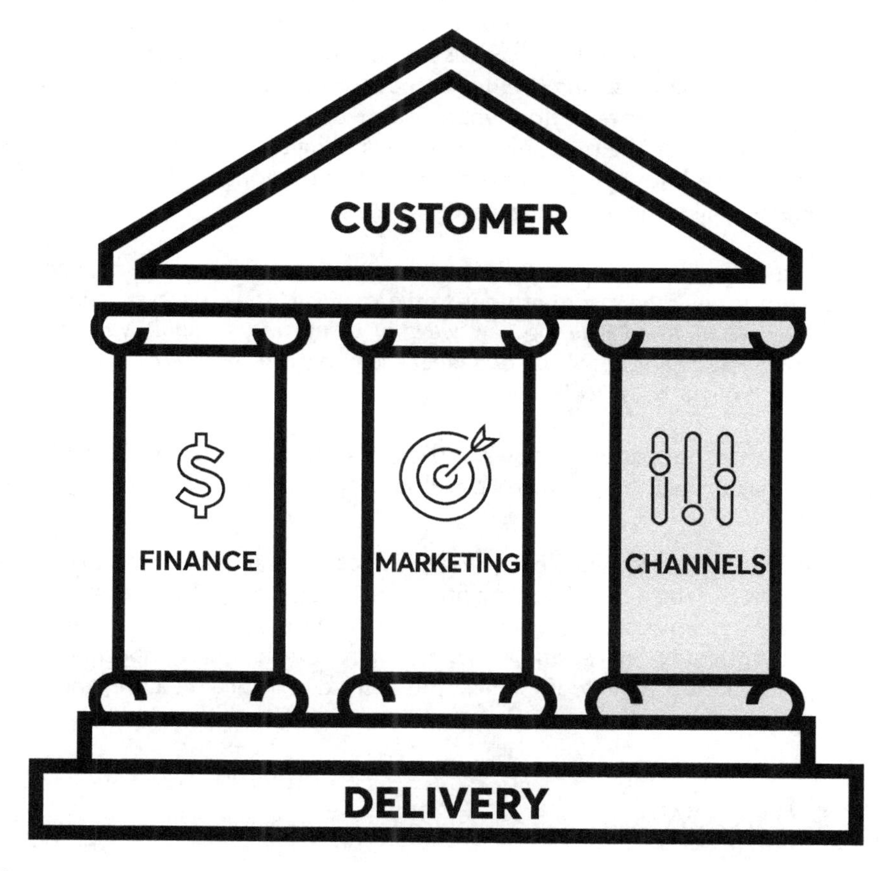

Figure 1-6. Alarm for channels

Regardless of how and where you first engage with your target executive, you have one shot at hooking them with a succinct statement that motivates them to pay attention and makes them want to hear more. Once you have developed this intriguing statement, consider it an asset that you can use throughout your business executive engagements, at any time: during phone calls, in e-mails, during future meetings or events, or even in impromptu conversations on an elevator or in the hallway. Your impact statement will carefully explain this in a way that motivates the executives to have a sense of urgency to learn more and meet with you. See the quick reference conversation guide in Appendix A.

Planting the Seeds of Expectations

After you have the next meeting set, let them know it will entail thinking differently and discussing how things might need to change. As soon as you have secured the meeting, the executive will want to move on to other things. You are safe for a few more seconds to ask a last question. Use this question to set expectations that you will be discussing the sensitive topic of changing the way they do things during the next meeting, which will become the seeds you will grow into a conversation when you meet. Frame your question in the spirit of preparing for the meeting by understanding their views ahead of time. Let the executive know you will prepare for this meeting by asking an open-ended question like the following:

> *Just so that I can best prepare for our meeting, is there a particular concern I should be aware of prior to the meeting? If we uncover issues, are you open to considering new ways of doing things? Could you tell me who else might be joining and what their concerns or interests about changing might be as well?*

Be sure to write these answers down and look up each participant's background. This information will help you to develop a conversation-starter for when you meet.

In the study "Strategy, Not Technology, Drives Digital Transformation," the authors point out the necessity of adapting quickly to change: "The 21st century is about agility, adjustment, adaptation and creating new opportunities."[5] A more recent study, "Increasing Agility to Fuel Growth and Competitiveness," similarly concludes that "companies must have a more flexible and agile operating model so the company can act."[6]

Let's Review

- Understand how your company is dealing with company and competitive pressures
- Research and identify the company problem and the ramifications of status quo

---

[5]Gerald C. Kane, et al., "Strategy, Not Technology, Drives Digital Transformation," MIT Sloan Management Review, July 14, 2015, https://sloanreview.mit.edu/projects/strategy-drives-digital-transformation/.

[6]Kris Timmermans and Donniel Schulmann, "Increasing Agility to Fuel Growth and Competitiveness," Accenture, 2006, www.accenture.com/t00010101T000000Z__w__/au-en/_acnmedia/PDF-4/Accenture-Strategy-Increasing-Agility-to-Fuel-Growth-and-Competitiveness-Research-v2.pdf.

- Be clear on the various alarms going off, how the landscape is changing, and how the executives are impacted
- Craft an impact statement that alerts the executive to those things that are not okay and stimulate curiosity and a sense of urgency to address it
- Plant seeds to manage expectations that the next meeting will entail a collaborative conversation

CHAPTER

2

Look in the Mirror

When the executive realizes there is a problem but hasn't yet realized they are part of the problem, it's time for them to look in the mirror. This is a critical conversation to bring them to that realization while respecting their views.

Designing the Conversation

The mindset of the executives at this stage is awareness that the company is dealing with a problem of increased competition and customer pressures, but lack of *knowledge* that they are a part of the problem. Whether you have created executive awareness that this company problem exists by sounding the alarm, or they had that awareness already, you must now give them the knowledge that their operating model is a contributing cause to the company problem. What's needed now is for the executives to understand how their current ways of operating are contributing to the company problem, and how they are important as a part of the solution. If they stay focused on their department silo and don't change perspective to a company view, they will never understand the ramifications of their company-inhibiting operating model. Their old ways of doing things are hurting the company in today's digital age. So, it's time for them to "look in the mirror" and develop a clear understanding of how their behaviors have contributed to their dilemma as well as how they will be instrumental in creating the future.

J. Orvos, *Achieving Business Agility*, https://doi.org/10.1007/978-1-4842-3855-4_2

This "look-in-the-mirror" conversation is not a presentation about agile, but rather a collaborative discussion to create a solution to the agreed-upon company problem. This conversation is designed to solicit their input and encourage them to share their own views, thoughts, and opinions about how to solve the company problem. As a result, the executive will be reassured that their perspectives are respected, and they will have input on any changes that would be required, as opposed to having change imposed upon them.

To conduct this delicate conversation effectively, you can follow these guideposts:

Set the Tone

In the initial moments of your meeting, the business executives are carefully observing how you come across and will make quick judgments about you. Business executives may be wondering, "Are you looking out for me and my goals or are you trying to have your way?"

Take control and set the proper tone in an elegant way by stating the purpose of the meeting: assessing the company problem (the one you shared in your impact statement), their perspectives, and how their operating model is a part of the solution.

In order to establish a collaborative tone from the start, you might say:

> *I would like to confirm that the purpose of today's meeting is to discuss some findings from new sources about the company's increasing competitive and customer pressures and to determine how these findings relate to how you conduct your business. Once we understand this, we can decide together if there are any problems worth solving. Therefore, I will be asking questions to better understand your current operating model and, based on what we learn together, we can mutually determine if and how best to proceed. How does this sound?*

Prime the Pump: Agree on the Problem

Because this conversation is about understanding the executive's operating model rather than making your own perspective understood, your role is primarily to ask questions and actively listen in order to allow the executive to come to their own conclusion. You need to walk a fine line of uncovering the gaps of their operating model without making the executive feel like you're judging, presumptuous, angling or manipulating, or wasting their time.

Handling these conversations as a collaboration will build trust. Start by focusing on business-related facts you can agree on—that digital disruptors are encroaching on your company. You will use this company problem agreement as momentum to get the conversation going. Open by sharing your research on this company problem based on insights you learned in the field, analyzing the market competitors, and listening to customers. Remind the executive that you are interested in their particular views and insights to solve the agreed-upon company problem as it relates to their operating model. In due time, as this look-in-the-mirror conversation unfolds, you will lead the executive to explore underlying causes, and eventually to consider that his or her own operating model is a key part of the problem.

For example, you might say:

> *We can agree that large organizations like ours are not structured today to compete against emerging agile disruptors and their ability to both sense and respond to market opportunities. As shown in the study Increasing Agility to Fuel Growth and Competitiveness (Accenture 2016), the majority of today's organizations have an inflexible operating model. "Only 24% of companies have a flexible operating model that can adapt to consistently deliver." Conversely, large organizations like ours are set up to support only our respective business unit silos with our own focus and goals. It is becoming clear that we will be at risk in the same ways if we don't think differently. Agreed?*

The prime-the-pump statement would look something like this:

> *As agreed, we all know that our customers are changing. (For instance ...). At the same time, competitors are emerging (such as...) and making moves to attract our customers (such as...). Our company may be taken by surprise and caught off guard if it doesn't respond quickly. (company problem).*

Then prime the pump with delivery, finance, marketing, and channel executives (Figures 2-1 to 2-4).

"Our **competitors** deliver software that customers are asking for more quickly than we can because we lack participation and input from the business."

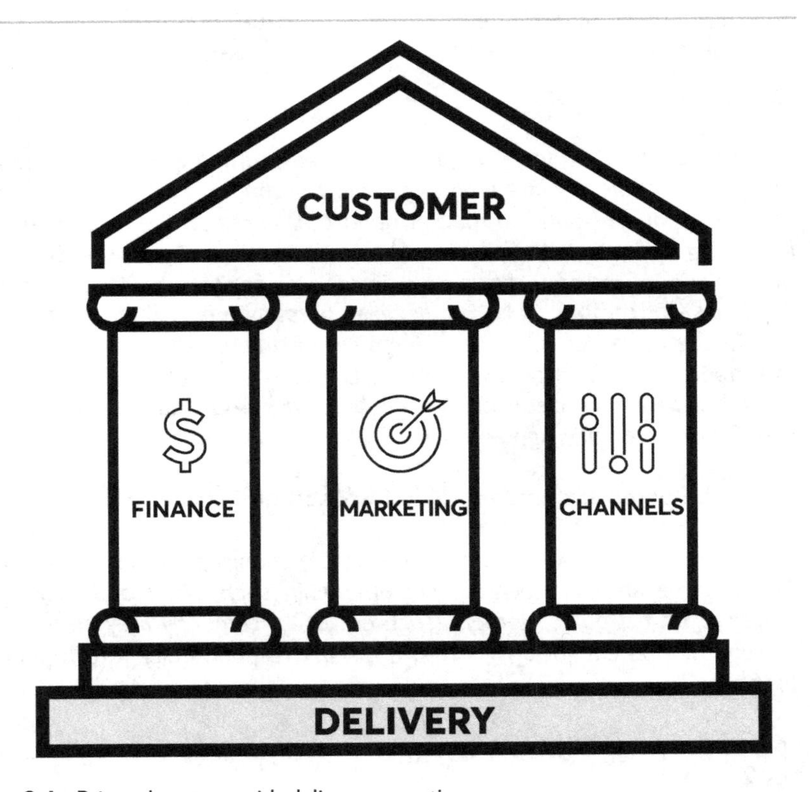

Figure 2-1. Prime the pump with delivery executives

"Our **competitors** financed products with significantly higher profit margins than ours, which is hurting our financial health in comparison."

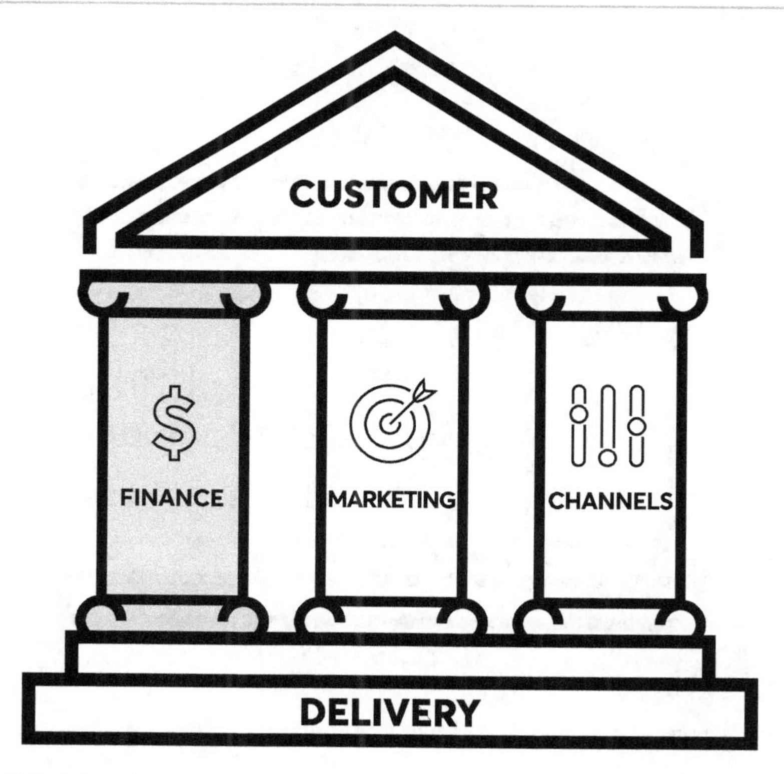

Figure 2-2. Prime the pump with finance executives

"Our **competitors** are attracting high-potential customers that we are missing because our messaging is missing the mark."

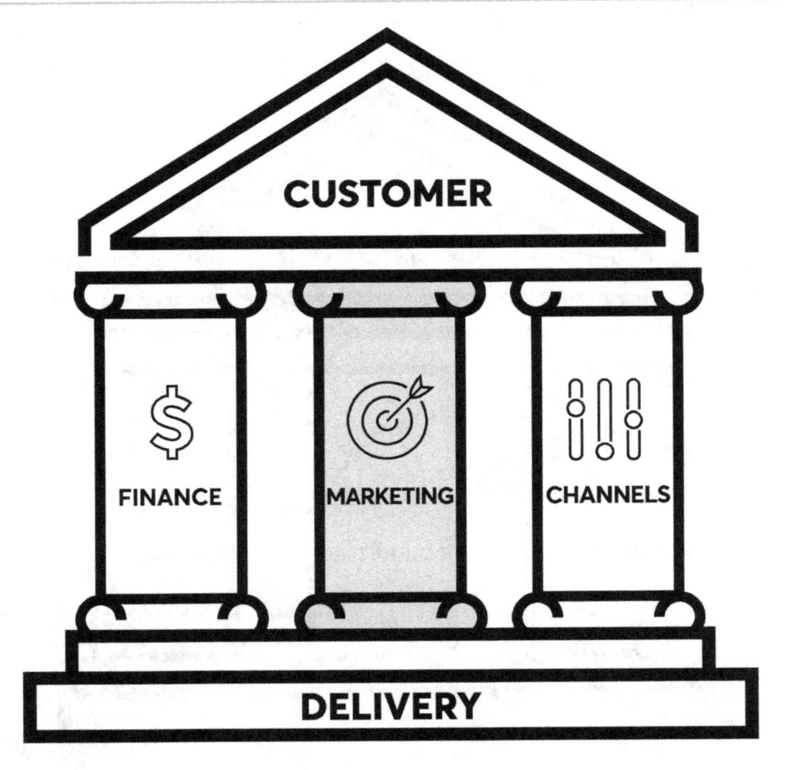

Figure 2-3. Prime the pump with marketing executives

"Our **competitors** are outselling our company in areas we are struggling in because customers see our channels as obsolete in comparison."

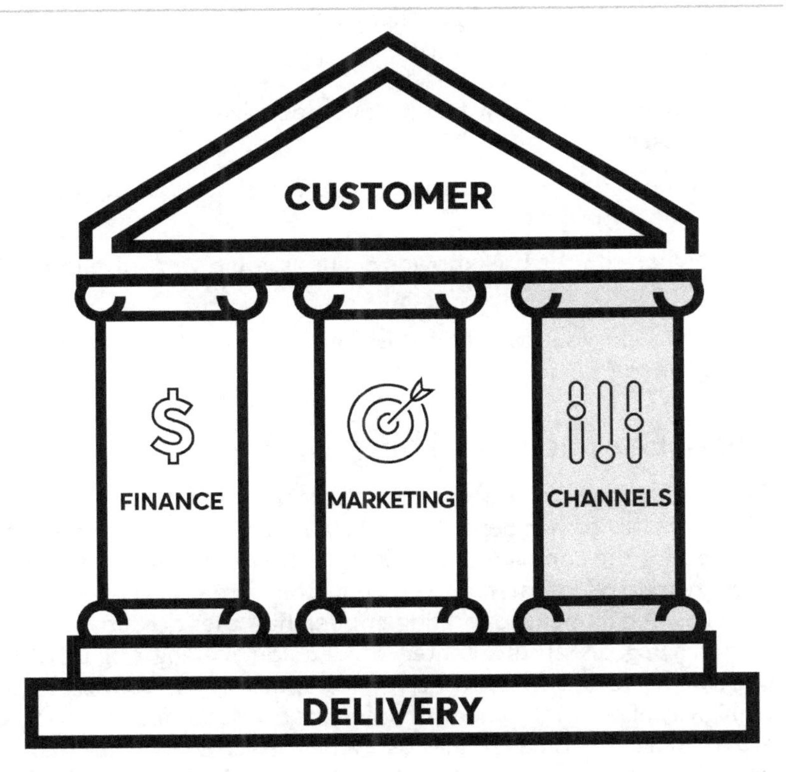

Figure 2-4. Prime the pump with channel executives

Your prime the pump statement is assumptive in that it concludes that you and the executive are already in agreement. Of course, if they disagree, or say no, then you need to stop the meeting and go back to sounding the alarm, since they lack awareness, which is required prior to any look-in-the-mirror conversations.

Elaborate on the Problem

Build on your initial agreement on the problem; let it simmer by asking the executive's views early and to get them talking about this agreed company problem. For example, you might ask:

- "What is currently being done to solve the company problem?"
- "When do you encounter obstacles to solving the company problem?"
- "Where do the breakdowns occur as you try to solve the company problem?"
- "Why do you do things this way to solve the company problem?"

Ask How Do They Do Things?

Shift the conversation to validate the operating model that the executive currently uses in his or her department. Establish that executive decisions are based on meeting the company and departmental goals—the company, and the executives' respective departments—are at the core for executive activities. This company/department operating model has the executive striving to achieve internal goals at the center of decision making (Figure 2-5). This translates to the four customer-facing pillars: delivery is striving to achieve on-time, high-quality development of software products, finance is looking to find ways to improve profits (saving costs and increasing revenue), marketing is working to attract potential sales leads, and channels is selling and supporting the company's products in the marketplace.

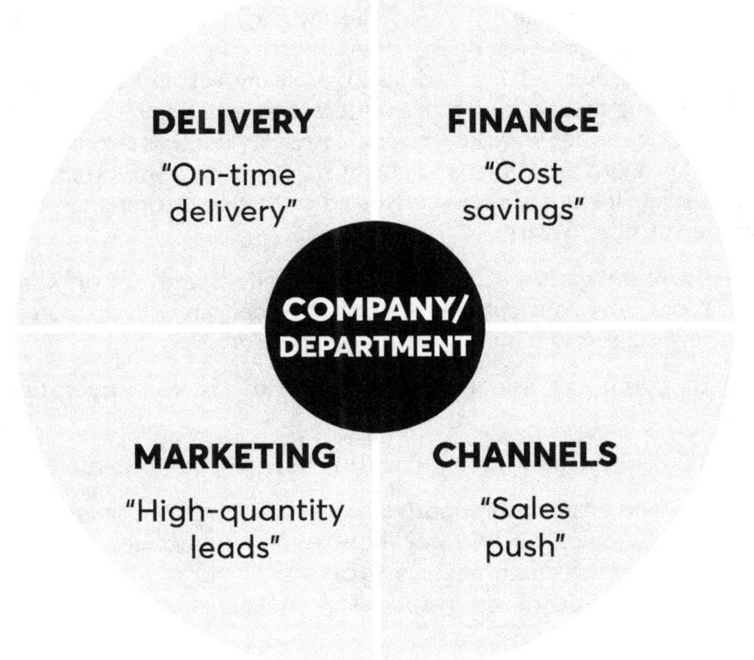

Figure 2-5. Company/department-centered operating model focuses on internal goals

Since some of the executives in these conversations may have supported or even created the company/department operating model that is currently in place, be respectful during this early stage of the conversation to establish trust. It's imperative that the executives trust you before you begin to reveal the shortcomings of the operating model. Frame this part of the conversation by acknowledging their company's/department's operating model decisions were made wisely based on the business environment at that time. Then point out that today's environment is completely different because of fast-moving digital disruptors and the demanding needs of customers. Be respectful and listen to build trust here.

Bite Your Lip: Let Them Own It!

You know that the solution to the company problem agreed to during the prime the pump conversation is business agility, and the ability for the company to sense and respond to the market. So how is the executive, using the company/department operating model, able to do this? You also know the answer to that question: it cannot be supported in the current operating model. Even though you know this, you still need to bite your lip to allow the executive to come to their own conclusion. Lead the executive down this path this by asking a series of open-ended questions that allow them to reflect and share their views. Asking and seeking to understand their perspective, rather than telling and seeking to be understood, will begin to reveal the gaps in the operating model without coming across as confrontational, judging, or presumptuous. With a positive and curious tone, use the open-ended "tell me more" approach, inviting the executives to elaborate, rather than responding with simple yes/no answers.

For example, you might say "Can you help me understand…?" or "Can you tell me more about how your current operating model can help the company deal with changing customers and emerging competitors?"

Going further, you can ask questions like, "How is your operating model helping to…"

1. Meet the emerging and changing needs of customers?
2. Defend against competitive entry of emerging disruptive technologies by knowing how our products are being used by the customer? Can you stay ahead of new kinds of competitors by responding to changing customer needs?
3. Attract new customers that we don't currently serve? Do you know how we can provide value to these customers?

When the executive answers, they are taking the first step, dipping a toe in the water, recognizing that *they are a part of the problem*. By answering these questions, the executive is taking ownership and indirectly admitting that their operating model is a part of the problem. From here, you can encourage the business to extend its initial observations about the company problem.

Explore Together

As the executives begin to realize that they don't have answers to all of the questions, they may start to feel exposed or defensive. Therefore, it is important to approach executives during this part of the conversation with empathy by listening actively and intently, concentrating on what is being said and visibly take notes to make them feel like he or she has been heard. Nod to encourage executives to talk about how they do things while you take notes. Pay attention instead of planning your next statement. Do not get overzealous. When the executive says, "We feel we don't have an answer to that," do not jump right in to announce, "I have just the thing for you." Instead, seek to create a collaborative conversation that encourages the executive to share their views and come to realize their company/department operating model is part of the problem.

It's time to take out your mirror to have the executive admit that their operating model is part of the problem because they are making decisions based on their department goals and constraints instead of the needs of the customer. The trust and momentum you have built during this conversation enables you to carry out this tough part of the conversation: create constructive tension by challenging that the operating model as an inhibitor to the company's ability to sense and respond.

Be careful not to cast blame on them. Instead, blame the changing market conditions. In other words, focus the conversation on how their operating model is a part of the problem because the market has changed, rather than their poor operating model decision. Reiterate this by saying,

> *We understand that current operating models were built upon the lessons of the past to solve the problems of the past. As new opportunities and challenges arise, our model should evolve to take advantage of new situations.*

In some cases, you will need further evidence to help the executives understand what it is that they are seeing in the mirror. There are several techniques that may help provide the information that the executive requires to help them get a clear picture of what is in the mirror. Listed below are some of these techniques and how they can help establish a sharper image in the mirror.

Face the Tough Reality: It's Time to Change

As you overlay this new digital disruptor environment on their current company/department operating model the gaps will simply begin to reveal themselves in the minds of the executives. Now that the executive feels like a deer in headlights, with questions they will have difficulty answering, give them an out.

Ask for views on how to improve their operating model, not presenting any solutions. This will give you their spoken words on the ideal solution, which will be important in the next strategy ("shine the lights," when you present the benefits of your agile business solution). Ask plainly, "What is your vision of an ideal operating model that can solve the company problem? If you had a brush and could paint the ideal operating model, how would that look? What operating model would ensure success in your view?" The company problem is the issue to be solved; do not dwell on the sins of their operating model, but rather focus on creating one that gets them answers to their challenges.

You are elbow-to-elbow in a collaborative approach, and your questions will encourage the business to "think out loud" about an ideal operating model and thereby hone their confidence in articulating it. Honoring the professional opinions of individual executives about the ideal operating model, ask ideal solution questions such as the following:

- "Please elaborate on how you would define improvement in this case."
- "Where are you currently in relation to where you want to be?"
- "How will you know you have been successful?"
- "What factors are contributing to the company problem?"
- "What will things look like when your ideal operating model is in place?"
- "What has worked or not worked so far to address this issue?"

When they answer these ideal solution questions, they are acknowledging their operating model needs to change to improve the situation. You are putting the mirror in front of them so they can see that the problem needs to be solved and they must be a part of the solution.

Delivery's Tough Reality

Given the progress and benefits of agile delivery, agilists may have a false sense of optimism that it can easily be applied to the business. That's because over the many years of practicing agile, delivery has been interfacing with product owners (or product managers) who give direction on what products and features must be built. Unfortunately, many product owners drive delivery based on a limited knowledge of what the customer wants or constraints in the operating model. There is an assumption that because delivery has been meeting their commitments and have scaled agile across the enterprise, that they have been building the right features, but it's not that simple. Once upon a time, this minimal interface into the "business" and the "customer" was enough. But looming disruptor competitors, with their finger on the pulse of the customer and able to deliver to their needs quickly, create a recipe for trouble.

COACHES CALL OUT: DON'T HAVE HAPPY EARS

Yvonne Kish Delaney, Sr. Principal Agile Consultant

A top US financial services company has over 6000 delivery employees successfully practicing enterprise "scaled" agile. As a result, delivery's high confidence has compelled them to lead the charge and expand their enterprise agile operating model into the business—business agility. Their global head of this effort said: "We are going to get all the pieces in place within delivery, and then we will be ready when the business is ready."

The assumption that the business will naturally become ready on their own is a very risky strategy. So to go ahead and paddle forward in delivery on a new operating model without collaboration or input from the business, early and often, can lead to several different things:

- Delivery could impose this operating model when the business does not agree to it. So IT creates a square peg for the business's triangle-shaped hole.
- Delivery can actively aggravate the business by creating this mountain of change without its input. Eventually, the result is a bigger rift between delivery and the business.

In other words, don't have "happy ears" and hear only what you want to hear by avoiding the tough conversations with your internal business stakeholders. Happy ears assumes that whatever awesome amazing operating model delivery builds, the business will naturally fall in line and come along for the ride.

Instead, delivery needs to shift towards embracing transparency and collaboration to engage the business early and hear their concerns or issues. Getting this bad news early might be difficult to hear, but it's the key to gaining business buy-in, building trust, and creating the right operating model.

Hearing the business feedback on your processes early is a courageous move because you might not hear what you want to hear. However, it saves wasting a massive amount of time and money spent on building an operating model that is ultimately rejected by the business. Just as early product input leads to powerful adjustments, early input on what the business likes and doesn't like in the new way of working can only help. Seek out feedback so you can understand what does and does not resonate. In the end, enabling the processes and techniques described in this book will greatly help everyone else to work towards adopting a new operating model. If delivery has maintained honest dialogue, when the time comes, others will be ready to consider learning from your lessons.

Sticking to this narrow product owner/manager dependency can inadvertently hinder a company from sensing and responding to the market, which of course, is the solution to the company problem. Ironically, product owner/manager dependency has given many agilists the false sense that the company is sensing and responding to the market. If they get the products out the door and the executives are happy, they succeed. From a business agility perspective, this myopic view and false sense of success for delivery is dangerous and can lead to their demise while agile disruptors swoop in and eat your lunch. When competing against digital disruptors, if your company keeps sticking to this traditional delivery-product owner interdependency, you will waste delivery development time and resources building software the customer doesn't use. That's because there's only so much a product owner will know about the customer (sense) and so much impact they can have on customer use (respond). They are taking the company to battle agile business digital disruptors while "sensing" with one eye covered and "responding" with one arm tied behind their back.

1. "Semi-sense" results in delivery building the wrong thing: products that customers don't value in the right way. That's because, in its current delivery-only agile operating model, there is sole dependence on a few people to make the decisions on what to build; these people are called the product owners (or product leads or product managers). Within delivery, these product owners focus on sensing what customers want in order to "build the right things." When customer demands are stable, predictable, and manageable, the product owner has usually been able to get this right with some success. Those were the good old days, but not the reality today.

   In the meantime, delivery is simultaneously focused on improving ways of "building things right" (agile) to make their internal executives happy. But while delivery executives claim success internally, the product owners might get it wrong and customers may not find value in those products being delivered—it's a roll of the dice. That's because delivery depends solely on the decisions based on the opinion of solid product owners, which is risky in today's fiercely competitive marketplace with fast-changing customer demands.

   Today, with the volatility of customer desires changing constantly, these product owners may not be fully in sync with their rapidly changing demands. How can delivery really be sure product owners know the right thing to build? They can't. As a result, delivery, while interfacing with product owners that may be out of touch with the current fast-changing customer desires, are building the wrong features that customers don't necessarily value. Just because your company has applied agile to Delivery, your company might be *building the wrong things the right way*. In other words, as one banking executive put it, "We know how to build things, but we don't know what to build."

2. "Semi-respond" results in building the right thing, but too late. When delivery builds something fantastic that the customer will love, they need to deploy it fast. It's very exciting to know when delivery has created a breakthrough innovation that customers have been asking for. But that's only one-fourth of the answer. There are three other entities that need to take this product to market and get the customer to use it. More importantly, and often forgotten, finance, marketing, and channels need to take the product to market. Once it's ready to deploy, these products then need to be priced, marketed, and sold.

Although delivery is already agile, it may be missing the mark in delivering products that customers can use or want, which results in low adoption. In fact, your customers may have already selected a competitor's product. Product owners have to face the new times with rapidly changing customer desires and product go-to-market demands to capitalize on small window of opportunity.

Delivery needs to look in the mirror and realize, in this new digital era, they have a high-risk agile operating model that has them creating products that today's customers don't want, or do want but receive it long after a competitive offering. Company success depends on the ability to create products that the customer values and can use on time. It's about the entire organization improving its ability to build "the right things at the right time," which depends on a better relationship with business. Success goes beyond agile in delivery and must extend into the business at a new level.

Today, delivery must form closer relationships with the customer-facing pillars to build the right products and to get these products to market. Without forming closer relationships with finance, marketing, and channels, delivery will continue to be at risk of building products that customers don't use either because their voice was not heard or because it's too late and they found other alternatives.

THE IMMINENT CHANGE AHEAD FOR DELIVERY

The way delivery executives are being measured for success is changing. It's frankly becoming "so 2000s" to be measured by delivering on-time and on budget.

Modern organizations will not be measured on metrics such as on-time or on-budget or on velocity or throughput. Instead, everyone and every team will be measured the same way, on the only metric that really matters: business results. This is refreshing for executives, as they have always been responsible for business results, but few others in the organization were. Now, organizations can have alignment on metrics from top to bottom. As a result, executive key performance indicators (KPIs) are evolving and changing with the increasing pressure to compete in a digital marketplace, and executives need to prepare in advance.

Even after 15 years of delivery agility, the focus has been on getting products built and delivered—build things fast and deliver them. But yet these products are not being used or valued by the customer. Since the 2002 study (The Standish Group "Chaos" report, 2002) that called out that 64% of software being delivered was never used, there has been very little proven progress. And even still, executives have been purely measured on getting products out the door. Agile practices just helped to speed up the process or work, since executives only had to make sure they delivered products the right way, regardless of whether anyone used them or not. But, again, times are changing.

In my travels with enterprise organizations it occurred to me that executives now are coming to us looking a bit lost. I would expect the opposite since their delivery group has been agile for years. These business executives are being told they need to be agile (not just delivery), they need to deliver more quickly, fund more frequently, plan more frequently. They are told all of this stuff through conferences, webinars, and so on. But no one is helping them get there. And it is overwhelming.

We are already seeing organizations change how executives are being measured, based on delivering customer value. Is the software they make actually being used? How can they plan for and then measure customer value? How can they understand when a product has delivered to its potential and therefore focus on other opportunities? Can they reliably understand (or enable) a product to hit the market first or while there's demand? How quickly can they change or create a plan? When the market changes, KPIs will be changing and executives will be measured by how the customers (internal or external) are actually using the product. So, cranking out a product no one wants, on time and on budget, will no longer be the measurement of success.

Delivery needs to prepare for the future. The future is building things right, but also building the right things. So, measuring executive success means you need to have a product that is shown to provide value. It not about reducing bugs, or being on time, it's now about holding executives accountable for the value they create for the company.

One important note. Product owners don't go away, but rather they become more important in this new operating mode (more on that in the next chapter). Delivery dependence shifting from the product owner to the new business relationships with finance, marketing, and channels will be a key element in a new modern operating model, which will be outlined in the chapters to come. In this agile business operating model, the product owner, as we know it, will become more important, not less.

CASE STUDY: "DELIVERY AGILE" IS MISSING THE MARK

A market-leading multibillion-dollar payroll company adopted agile in delivery and has many product owners and product managers giving direction on what to build. They established an "innovation center" with several hundred software developers and product owners using scaled agile ways of working. They were building products based on the direction of the views of these product owners. Sometimes they got it right and sometimes they missed what the customer wanted. It was random and difficult to predict what the customer wanted, since the entire product decision was based on the product owner's options. They didn't have real customer voice intelligence from marketing or channels.

But then they got it right. A potential breakthrough innovation. However, their innovation center was solely dependent on the product owner, and not connected and aligned with the customer-facing pillars that will take this product to market. So, when delivery was ready to deploy that software innovation in June 2016, the other customer-facing pillars were not prepared to support this because they were not involved with the planning. Finance didn't know how to price it, marketing didn't know how to create the right message, and channels didn't know how to sell it. As a result, the release was postponed by two months. And while the delay was happening, a digital disruptor launched an alternative innovation with the right price and marketing message, arriving first and capturing market share.

Finance's Tough Reality

Finance leaders need to look in the mirror and face the reality: finance has been funding products with low profit potential. That's because they are stuck having annual finance budgeting events that are set and do not change until the next year. As the finance department aims to optimize their spending with prudent product investments, these decisions are based on a best guess, since the annual events don't allow for change. When finance uses a rigid, outmoded annual budgeting that commits funds to products that have low value to the customer, the business is forced to set lower prices that yield lower profit margins. The funded products are, in turn, commoditized quickly by competitive innovations, which causes prices to be set at lower levels with low profit margins.

Helping finance to look in the mirror and see the constraints to their process means revealing how they have primarily served their own department, where success is often gauged by how efficiently they spend their budget. In this current model, finance is continually thinking of their funding strategies based on what the competition offers and trying to fund products that come close or perhaps offer a lower price, resulting in lower margins. The culminating problem is that their funding strategies will keep the business constrained to pricing product at the whim of the market.

Finance funding decisions, therefore, need an operating model to allow for these innovations to become unleashed. With better funding decisions to build products that customers demand, better prices can be set by the business.

Marketing's Tough Reality

Marketing reaches *customers* by creating campaigns to pique interest in new products. The current operating model aims to implement campaigns that generate as many leads as possible, which then can be converted to opportunities and potential revenue. In our competitive digital world, the difficulty arises when the company's marketing messages are constantly a step behind those of competitors.

The executives in this department need to see that they are promoting things that have already been adopted in the market and overshadowed by competitive offerings. The customers they are trying to attract are being bombarded by marketing messages from all sides simultaneously. In the crowded waters of this bloody "red ocean," with so many sharks going after the same bait, a rigid marketing model doesn't stand a chance. In reality, it leads to campaigns that generate a large number of low-quality leads from unmotivated buyers, ultimately resulting in low sales conversion.

Your look-in-the-mirror conversation can help the marketing executives see that their traditional operating model is cutting off their access to many higher-quality leads for the company.

Channels' Tough Reality

Channels is the customer-facing pillar that focuses on how and where to sell, support, or implement the current product offering. Customer interactions occur through two primary communication mediums:

1. Customer service associates, consultants, and salespeople on the phone, in the field, or on the retail floor can serve as the customer's access point for the company product. Salespeople in the field or in a store sell inside telesales. Sales and customer services field forces who engage

customers directly have an intimate understanding of what they want, so they are most successful in selling the company's products and services while only being capable of selling what they have to offer.

2. E-commerce software or mobile apps can also serve as the connection between company and customer. For example, LightStream, an online consumer bank (division of SunTrust Bank), provides customers with good credit scores, a simple and fast loan process, and excellent rates through a few clicks of their mobile app. This channel is fast evolving to include Internet of Things, like having your Amazon Alexa or Apple Siri integrate with the customer's account: "Hey Alexa, get me a rate quote for a car loan!"

Although the channels pillar has an intimate understanding of what products are most successfully selling and what customers want, their hands are tied and they can only sell what the company has to offer. As a result, channels relies on pushing whatever product features and functions exist onto the customer without being afforded the opportunity to guide have input and influence on any improvements or ideas.

When you ask about their ideal solution with an altruistic mindset that puts the business executives at the center of attention, your credibility can skyrocket. You allow the executive to envision and elaborate on ways of improving his or her situation. This is an assumptive close for the executive to indirectly admit they are a part of the problem and motivates them to take action. By encouraging the executive to speak freely and think out loud about an ideal solution, they articulate their opinions, thoughts, and views. This helps them become more aware and enlightened. Simply write their spoken words under the heading "Ideal Solution." You will have the opportunity to respond when you use these spoken words as your key asset during your next step when you "shine the light" (see the next chapter).

Let's Review

- Focus the conversation on the executive, not agile
- Prime the pump to earn the right to ask questions
- Create a collaborative conversation by asking how the executive does things
- Create tension by addressing operating model constraints
- Get them to state their "ideal solution," which enlists their participation as a part of the solution

CHAPTER

3

Shine the Light

You can shine the light on a new way to make product decisions that will best deliver customer value. Your task at this stage is to demonstrate how the executives from the four pillars, delivery, finance, marketing, and channels, can benefit from a new operating model to address the company problem. You will, therefore, need to take action as the first step toward this change by illuminating how each executive will benefit by improving their operating model to be solve the company problem.

Presenting a New Operating Model

Use the shine-the-light strategy when an executive acknowledges, admits, and accepts that their company/department operating model needs to change in order to address the company problem. They are open to looking at alternative operating models to solve this company problem. Although executives at this stage may be open to change, they don't necessarily know how. So that's your job now. It's now time you give them the answer they are seeking. Your shine-the-light conversation will show them a new improved operating model that will solve the company problem and benefit them as well. It's time to shine the light on the solution, giving the executive the *ability* to implement a new improved operating model. When you shine the light, your action will create a tipping point for the company to take action.

J. Orvos, *Achieving Business Agility*, https://doi.org/10.1007/978-1-4842-3855-4_3

Creating Alignment Across the Pillars

As emphasized throughout this book, in order for the company to solve its problem, it must be able to sense and respond to deal with digital disruption threats or customer changes. That requires delivery, and its software development efforts, to become aligned with the other three customer-facing pillars: finance, marketing, and channels (Figure 3-1). These customer-facing business pillars will enable the company to sense the customer demands and then respond with a product go-to-market plan to support the products that delivery builds.

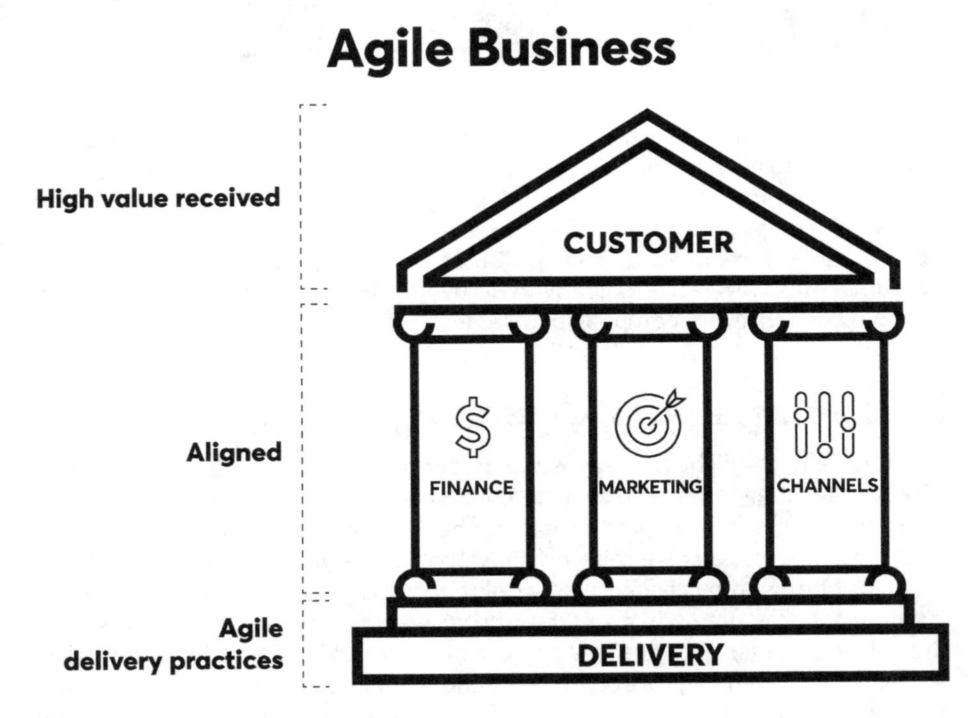

Figure 3-1. Alignment across the pillars

Today, in the digital disruptor world, a company cannot place its bet on delivery producing a product without having the customer voice (sense) and the complete product go-to-market plan (respond) from the other customer-facing pillars. Any software or software-enabled product or innovation that delivery builds must be based upon accurate input from these other pillars and then wrapped in a product go-to-market plan as well: finance needs to fund it, marketing needs to promote it, and channels need to sell it. Therefore, a product is much more than just making software in delivery. It's about the product being used by the customer, which creates a business result, which in turn requires a full product go-to-market plan. To reinforce this point, we will synonymously refer to products as product go-to-market (which includes product).

This operating model is built to change and will allow your company to deal with the company problem by

1. Meeting the fast-changing needs of customers by building products that they will truly value and establishing continuous feedback loops to help anticipate future interests.
2. Making product go-to-market decisions based on weighing the investment requirements vs. The potential profits from the value that is created for customers.
3. Defending against competitors and staying ahead of unexpected emerging "agile business" digital disruptors by better understanding how customers use your products.

Customer-Centricity Drives Alignment

In contrast to the outdated company/department operating model, in the new, modernized operating model, the respective departments will have a different focus: the customer. This **customer-centered operating model** enables a company to sense and respond and is the core of an agile business. Instead of focusing on the company and departments winning, the new focus is how the customer can win (Figure 3-2). This move is a shift away from conducting activities that help the company on the back of the customer and a shift toward helping the customer on the back of the company—that the company's purposes are to provide a base for holding up the customer and to provide value. As a result of this change in thinking to a customer-centered operating model, the respective executives can benefit in the following ways:

1. **Delivery** becomes the source of creating value by developing products that customers will use because it is designed with their needs in mind and launched to market before competitors.
2. **Finance** becomes the source of generating profit by making smarter software-funding decisions that keep up with fast-changing customer desires and open the door for the business to set higher price points.
3. **Marketing** becomes the source of more effective messaging to customers that are more likely to buy.
4. **Channels** become the source of software improvement ideas, and value feedback, serving as an antenna for customer feedback and the market.

Outcomes When the Customer Is at the Center

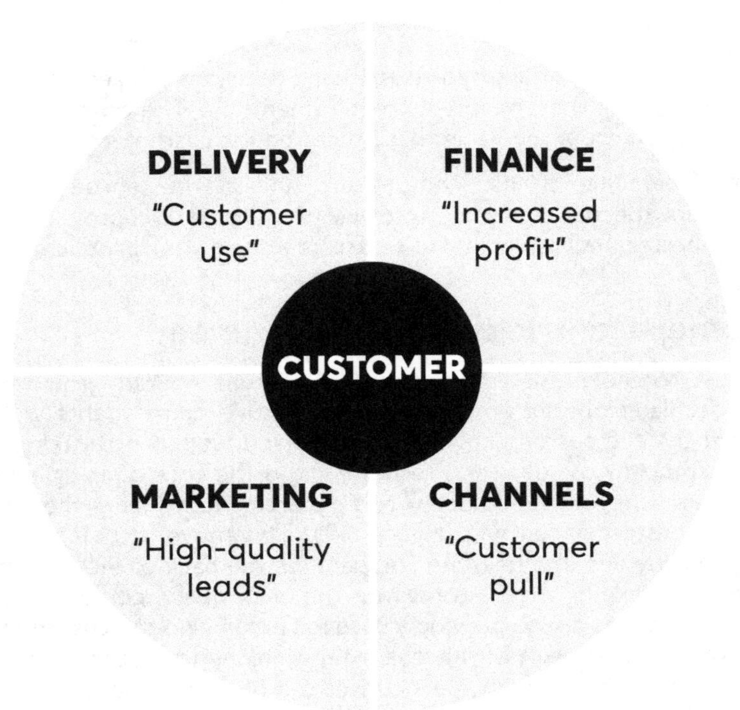

Figure 3-2. Outcomes when customer is at the center

By moving to a customer-centered operating model, the customer-facing pillars converge on customer needs and respond with consideration.

Communicate Business Benefits

Conduct a presentation that outlines the benefits that each business executive will gain by changing from their current company/department operating model (as discovered during the "look in the mirror"), to an agile business operating model, which is the new customer-centered operating model.

This presentation will start with the anchor of these two upfront agreements. Begin the conversation by establishing these agreements on the mutual understanding that

1. The company's current company/department operating model is not designed to handle the company problem. These could be based on the agreed executive knowledge and understanding, either from them or from you, during previous look-in-the-mirror discussions when you gave them this knowledge. Either way, this is the starting point of the presentation.
2. It is necessary to agree on the purpose of the meeting. Since these executives will need to take action by cocreating a solution, the purposes of this meeting are to present a new operating model that will solve this company problem and benefit each department and to determine whether the executives would like to conduct a pilot as the next course of action.

The presentation will continue by bringing in the executive's own spoken words about their ideal solution (as discussed in Chapter 2). These spoken words are are key because now you will be able to present solutions using their own words and expressions. By doing this, you continue to build trust with the executive to show them you are listening and are taking their views seriously and incorporating it into what you are about to present: the **customer-centered operating model**. Sharing this using the customer's own words will reduce the sting and eliminate any major resistance. After all, you are basing this model on some of their own ideal solutions.

What's in It for Delivery?

With a customer-centered operating model and the new focus on customer value, delivery will aspire to aim to "build the right things"—products that customers are asking for and will use—rather than merely taking direction from product owner proxies to deliver potentially stale products on time (Figure 3-3).

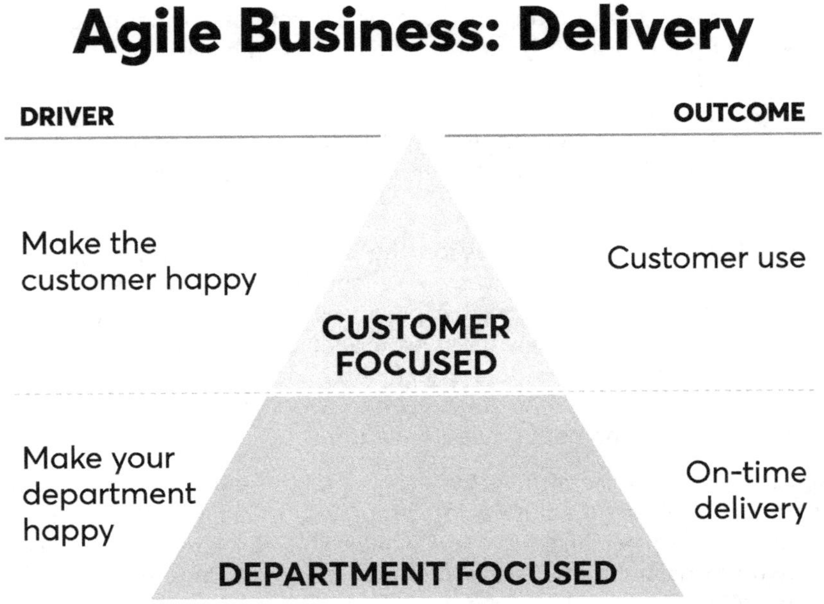

Agile Business: Delivery

DRIVER | OUTCOME

Make the customer happy

CUSTOMER FOCUSED

Customer use

Make your department happy

DEPARTMENT FOCUSED

On-time delivery

Figure 3-3. Agile business: delivery

By working as a team with other pillars in a customer-centered operating model, delivery can double down on its current agile delivery practices. Not only will delivery have the ability to develop new products in a timely fashion, but they'll also be able to go to market faster instead of getting held up by rigid external business barriers. In essence, they will be successful by embracing the new operating model.

When delivery knows that what they build will actually be in the customer's hands to use, it will be an exhilarating and motivating feeling for the staff. As a result, delivery can expect a resurgence of employee morale and engagement, which helps drive to being in time to meet the customer opportunity and with the higher quality customers have come to expect.

What's in It for Finance?

By making customer value the focus in a customer-centered operating model, rather than just cost savings, finance can ensure they invest in the right things, which will increase profits and shareholder value (Figure 3-4).

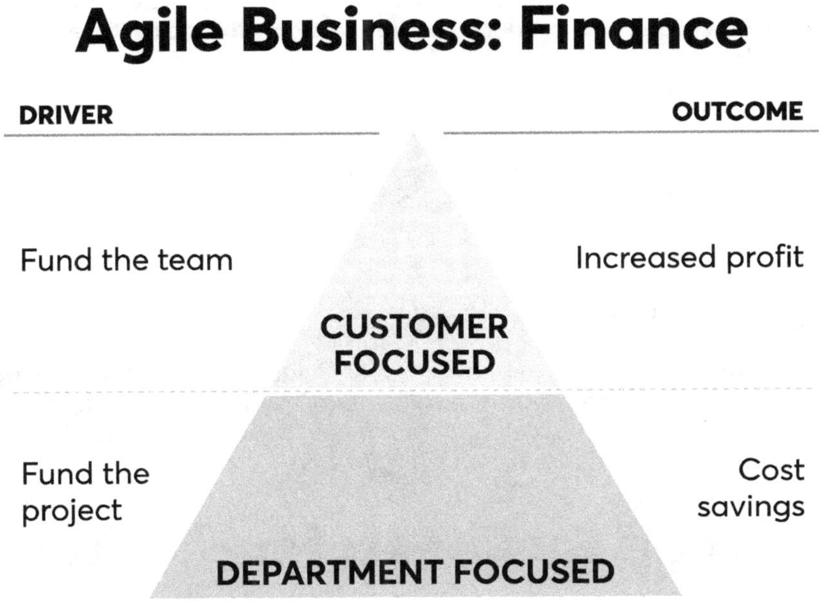

Figure 3-4. Agile business: finance

First, finance should fund products and features for which customers are willing to pay a premium. The results will be increased profits because the business can set prices higher.

Second, finance will spend more wisely and avoid wasting money on efforts that will not deliver the highest value to customers. Since marketing and channels will provide a current and clear understanding of what the customer actually wants and needs, finance can make better funding decisions and become confident they are investing in the right products. To further this, delivery will have and be able to provide transparency for finance. Finance will shift their focus of funding projects in the company/department operating model to now funding delivery teams that develop the product. By funding delivery teams, there is a much higher level of transparency and visibility into the work being done, and therefore the potential customer value. By working with delivery, marketing, and channels, finance will get the most value for their investments and can be confident that expenditures will fund teams to deliver the capabilities that customers value.

Finally, finance reduces risk because they are making fast and small investment decisions to properly react to the market. Long-term, annual, upfront funding is eliminated and replaced by a well-informed, incremental, small-outlay funding approach based on customer value. Therefore, finance does can make smaller, more informed bets on product changes instead of funding based on gut instincts alone.

With a dynamic investment approach, they can reduce the risk of products that don't perform in the market by being able to change their funding decisions to reallocate their resources to deliver the best products. Therefore, finance will benefit significantly from business agility since it can positively affect funding decisions.

What's in It for Marketing?

With customer value at the center of product go-to-market campaign decisions, marketing "promotes the right things" and makes messaging more relevant to customers in order to improve the quality of potential leads and increase the conversion rate to sales (Figure 3-5).

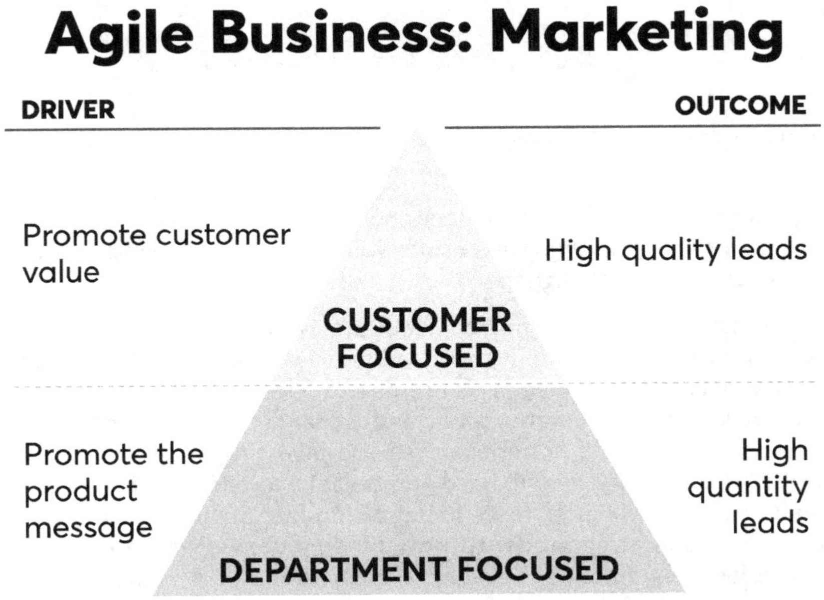

Agile Business: Marketing

DRIVER
OUTCOME
Promote customer value
High quality leads
CUSTOMER FOCUSED
Promote the product message
High quantity leads
DEPARTMENT FOCUSED

Figure 3-5. Agile business: marketing

The customer-centered operating model emphasizes gathering and using customer data to generate messaging that resonates and that provides value on its own, in advance of product.

First, as campaigns provide messaging that resonates more with leads and what they are asking for, marketing can expect to cut through advertising noise with their own distinct message, increase customer web traffic, and bring in quality leads that can convert to opportunities and sales measurable in revenue.

Second, marketing's contribution to the bottom line will improve because they are helping to drive demand toward the higher-margin products rather than trying to stimulate demand in commoditized and lower-margin product offerings.

With today's demanding audience and fast-changing tastes, marketing needs to create fresh messages in the eyes of the customer. Drawing upon input from the other pillars in an agile business process will give them privileged, up-to-date insights. Furthermore, by planning and delivering in shorter cycles, all the while gathering customer feedback, marketing can hone in on what messaging resonates and what does not. Keeping the message current and in line with the new and evolving product offerings will help ensure the brand is still viewed in a positive and successful light.

The benefit of the marketing team collaborating closely with delivery, finance, and channels is to help increase its own effectiveness by ensuring they are promoting only those products that will be delivered. In the old company/ department operating model, marketing may have had false starts when delivery projects got cut or changed. In the old model, when product changes occur, marketing was unaware and wasted effort. But in the new world, marketing can focus on the work that aligns with the rest of the pillars, ensuring the most effective approach.

What's in It for Channels?

By emphasizing customer value in decision-making in a customer-centered operating model, channels influences the software product by leveraging its understanding of the customer journey, perspectives, and feedback (Figure 3-6). This customer voice can now be heard by the company and no longer fall on deaf ears. Channels can provide these customer insights to a company that will listen and use it as fuel to make the right product decisions.

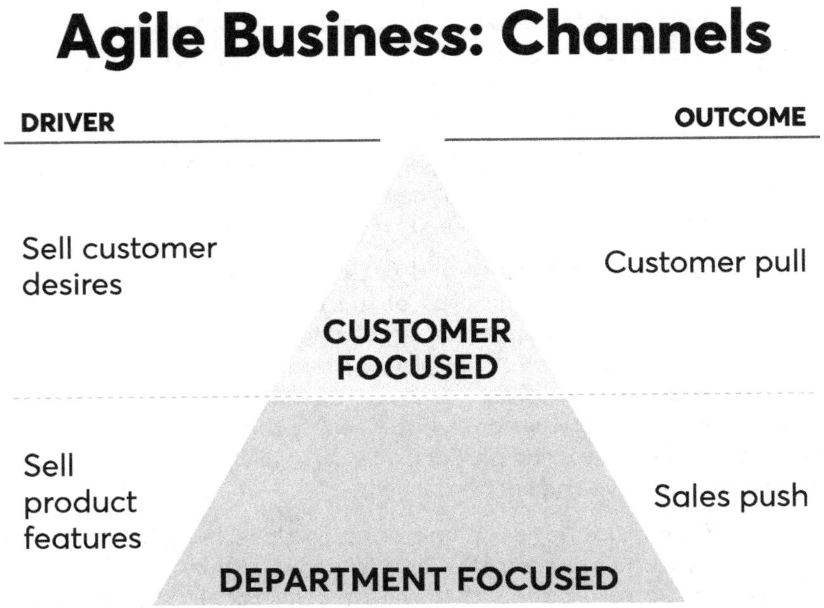

Agile Business: Channels

DRIVER

OUTCOME

Sell customer desires

Customer pull

CUSTOMER FOCUSED

Sell product features

Sales push

DEPARTMENT FOCUSED

Figure 3-6. Agile business: channels

Channels is constantly interacting and intimately working the customer. Channels retains a wealth of information about customer desires based on engaging with them directly multiple times. However, this important customer intelligence is unfortunately not shared with delivery so they can make better products. This is a waste of all the amazing information we gather while listening to the customers on a daily basis.

As an antenna to the marketplace that can detect changing customer desires and requests, channels can now effectively communicate opportunities or competitive threats, and if necessary, the company can pivot. Channels also increases customer intimacy by serving customers with relevant product offerings that meet their needs and wants. This pillar now becomes a means for customer interaction and provide valuable feedback about what they are looking for and your company is no longer confined to just pushing products and hope for adoption. Upgrading the old one-way product-push to a two-way responsive interaction will ensure that customers are heard and validated, and the sales channel will gain the ability to increase sales.

The channel is an excellent way to hear the voice of the customer, which can be used for the company to build better products. Channels would relish the opportunity to share their insights to positively impact the delivery process if given the opportunity.

When channels can connect with customers in a more meaningful and intimate way, that's power. In the new customer-centered operating model, channels will become closer to the customer and ultimately attract more customers since they can effectively address their ideas and requests for product improvements. The company will impress customers by proving that they have tuned in to requests to enhance the product.

Moreover, channels can genuinely welcome customer requests, or even complaints about the product, because they have a responsive, pivot-ready agile business team backing them up now.

Channels will, in turn, be the root of all customer interest intelligence in the feedback loop to the other departments. It's up to channels to share any blockers when a customer is looking for something that the product lacks. And channels can proactively ask customers questions like these: What do you love about our product? What's wrong with our product? What can be better? What do you want more of/less of? The information gathered from this back-and-forth communication will enable channels to inform the other three departments of where a product improvement can have a direct impact on sales.

Coalition to Build Momentum

After your presentation, find the like-minded executives and assemble them together with the common understanding that they need to change the operating model. Once you identify the executives that are on board with the benefits of a customer-centered operating model, it's important to assemble a like-minded coalition.

To create groundswell support, you will coach your executive to become your internal communication partner who will facilitate conversations between other executives and create favorable office chatter. You will need to invest time and effort to coach and equip your champion to be successful. Provide them with assets they can use to influence their colleagues. Specifically, set them up with tools, including relevant articles, research conclusions, and websites, that will reference your conclusion that their process must change and improve.

Once they are armed with this information, ask your executive to join you in approaching their fellow executives from other key pillars (delivery, finance, marketing, and channels). Encourage and coach your executive to start a domino effect by asking that each one then share their favorable outlook to the others. Through these efforts and actions, opinions can grow into a groundswell, with executive stakeholders across the customer-facing pillars (delivery, finance, marketing, and channels) talking together in favor of improving their process.

COACH'S CALL-OUT: THE IMPORTANCE OF BUILDING A COALITION

Saif Islam, Agile Managing Consultant

In one of my engagements with an insurance company, 80% of my time is focused on building the coalition of support that there is a problem with their current operating model. Most of my efforts were around this area and conversations continued over weeks to ensure that the executives agree that there needs to be a change to an agile operating model that was designed around delivering great products in order to deal with company competitors taking their customers.

As for many other large enterprises, executives in different roles had different priorities, areas of focus, and success measures. Because the idea of developing customer-valued products was not clearly owned by anyone in delivery, marketing, channels, or finance, they didn't focus on this, so we had to change their orientation. We had to show them how they can benefit from delivering customer value. We had to shine the light on these benefits, which they had never considered.

After we told people about this new operation model, we created a guiding coalition of key executives from delivery and the other three pillars, marketing, channels, and finance, who all have "seen the light" of customer shining. Now we just wanted them to make sure they can be the shining light to communicate the benefits of the customer-centered operating model.

- They received training and coaching so that they could support their teams in the new way of working, leveraging agile principles and practices.
- It was important for them to understand overall transformation strategy, structure, and execution approach.
- It was also important for them to understand how to handle the organizational and cultural change journey of large cross-functional teams.

As a result of building this coalition, we created a buzz and some excitement. It helped them get a shared vision, which is a precondition for successful change. Having a strong guiding coalition with a shared vision can significantly increase the pilot's chances of success.

- A list of executives was identified in all four pillars (delivery, finance, marketing, and channels) whose leadership was needed in this initiative. There were one-on-one meetings with them to understand their pains/gaps, objectives, priorities, and success measures. There were additional conversations with key leaders in each of four pillars to understand the root causes of the pains/ gaps and their impacts.

- Then a workshop was conducted with the executives from all four pillars to form the guiding coalition, and to get to a shared vision.

Pilot to Take Action

Now that you have shined the light showing how delivering customer value can benefit the company, take action to prove these benefits with a pilot. Put these concepts to work on a real product, and generate real benefit for the customer. This pilot is an important step in the journey to becoming a responsive, pivot-ready organization. This pilot will become a tipping point or a major obstacle depending on the results.

The purpose of this pilot is to prove the benefits in your presentation by showing the results in a live demonstration in a short time frame (for example, two months). For the most impact, select a real program that leads to real, deployed software and generates real customer value that can be seen. You will showcase agile ways of working to become visible and tangible in a live demo after this short time frame. When others can see how great it looks in action, the impact will become a tipping point. When you show how it looks to deliver customer value quickly in a live demonstration, they will see the light you're shining.

Done right, the pilot will result in the following:

- Better communication among the customer-facing pillars: delivery, finance, marketing, and channels. Agile planning and delivery cycles make it easier to communicate what will be delivered, update on real progress, and share feedback from customers.
- The potential for better alignment among the four pillars. When delivery becomes something everyone can count on, then the four departments can have more productive conversations about what the right work is and how to support that work.
- A framework that will enable organizational pivots. The customer-centered operating model enables delivery to deliver products in a different way—features are delivered whole without a long tail of unfinished but required work. A well-designed pilot helps everyone see how the business could pivot painlessly after a release.

- The powerful result of delivering customer value. Most people in the company won't expect customers to be delighted with the smaller scope of a customer-centered operating model release, but usually customers respond enthusiastically to getting even a little of what they *want* in so little time.

COACH'S CALL-OUT: THE PILOT EARNS THE ABILITY TO TAKE ACTION

Ronica Roth, Advisor, Agility Practice Development

One leading online retailer ran an agile pilot on a new search functionality for their e-commerce site. Sounds simple, but this was many years ago and the #1 competitor was killing them on better search results. It was a high-profile project. Everyone was counting on the agile experiment helping them learn rapidly in an area where they had struggled mightily.

The team built and released functionality to a beta site every week. Half the company, it seemed, was getting the new code and playing with it, including the CEO. Channels and marketing were definitely paying attention.

The frequent demonstrations provided the delivery team with a ton of feedback, and it gave the business a real heads-up as to what was coming. The business began to coalesce around the possibilities of a better customer experience and how to leverage that. And agile got real support, now that the impact was material.

As a reminder, bringing in consultants when you are presenting the benefits is too early in the process, but now is the time and you can hire them to help deliver the pilot. Given the importance of conducting a quick live demo that shows the customer-centered operating model in action, the stakes are high and it's time to get help and bring in professionals.

In order to take action with a successful pilot, follow these two principles:

- Principle One – Establish a pilot team
- Principle Two – Decide on the minimum viable product (MVP)

Principle One – Establish a Pilot Team

The pilot requires a team of leaders from all four key areas of the business: delivery, finance, marketing, and channels. This is a pilot of the new customer-centered operating model for the business, not just a new way of building software. The rest of the business must be involved from the start. We need business leaders to guide the pilot to ensure that what we learn applies to the whole business and to provide backing and cover for trying new and different procedures.

Too often agile pilots are run from inside delivery, with no connection to the larger business. When that happens, the larger transformation starts a step behind—even a successful pilot will be seen as irrelevant, if it's seen at all. The champions will be left playing catch-up, trying to explain what's been going on for the last quarter and why anyone should care. Meanwhile, the successful delivery folks will fall into the same arrogance trap as those agile consultants—wondering why everyone doesn't beg to work like them. While they played with new ideas, the rest of the business was continuing the important work of guiding company success.

Participation

Approach the pilot not just as a test of agile delivery practices to deliver software, but also as a test of bringing together the whole organization in a new way of working. You do that first by recruiting a leadership team to guide the pilot, monitor progress, assess results, consider implications, and suggest next steps. That leadership team must include representatives from delivery, finance, marketing, and channels.

For the pilot, the cross-departmental leadership team will not necessarily comprise senior executives—at this point you need people who can be close enough to the details of the relatively small pilot to remove obstacles and keep things moving forward. Too senior, and the pilot will suffer from inattention. Too junior, and the team won't be able to remove the organizational obstacles to success (like excusing the pilot from standard processes or bringing together people from different departments). The answer will depend on how hierarchy and power work in your organization. Regardless of level, do be sure to include all four departments.

Within this pilot team, identify your sponsor executive; this is the leader who will be able to represent the goals of your pilot, ensure that each short-term target is achievable, and provide funding if needed along the way. Your pilot team may have to work very hard to come up with these targets, but each "win" that you produce can further motivate the entire company to join the effort.

A clear vision can help the team understand why you're asking them to do something different. That vision should be result-oriented, not activity-oriented. So, it's not about the activity to but rather the results to "show we can get to market sooner" or "improve the quality of the pilot release, as measured by post-production support tickets." And better yet if these results are being tied to a larger business strategy, like "gain market share" or "win against Competitor X." When executives see that you're trying to achieve results that benefit them, then the directives they're given tend to make more sense. What you do with your vision after you create it will determine your success.

Establish a peer team of delivery, finance, marketing, and channels as a trusted source of strength and direction, instilling confidence so these executives will support each other to achieve a common goal to have a successful pilot. Peer-level communication will allow everyone's perspectives to be considered before making decisions on what to build in the pilot.

The team sets a regular cadence of leadership team weekly meetings to plan how to support the pilot as a strategic initiative. These meetings require each executive to make an explicit commitment that has measures and specific actions attached to it. Then the team needs to meet on a frequent cadence—daily, or two or three times a week--to check in on that commitment, and to see who needs help/is blocked. So if the commitment is in danger, is there something the rest of the team can be helping with? How can they resolve this together?

Your message will probably have doubters within the company, so you'll need to communicate it frequently, even daily, with the pilot team. To communicate the pilot vision, talk about it during everyday conversations as well as the weekly team meeting. The team will use this pilot vision to make decisions and solve problems. When you keep the vision fresh on everyone's minds, they'll remember it and respond in a more informed way.

Since the peer communications are based on a common vision for the pilot, transparency becomes a conscious event when outlier ideas are discussed and handled openly as a group, rather than becoming underlying derailments. For example, if the channels executive is planning to initiate a new sales tactic that isn't related to the strategic initiative that delivery, finance, and marketing are tracking, there is moment of truth from these peers as they hold channels accountable to stay aligned with the common goals of a successful pilot.

COACH'S CALL-OUT: TRANSFORMING TAKES A TEAM

Christen Mclemore, Managing Consultant

I worked with a Senior Director of Operations in a Fortune 500 company and have been working on transforming their industrial business to become digital. They had attempted to transform their business and system processes a couple of times and each time they struggled to get their business to participate in the transformation team. While they believed the idea to participate was great, they couldn't attend because they were simply too busy.

The transformation team decided to move forward without them, selected a business unit to focus on, and kicked off our transformation team with people in engineering, QA, operations, and project management. They began by identifying the internal business challenges that were delaying deliveries and preventing them from meeting commitments. Not having the business represented was brought up several times as an impediment because many of the changes that were being presented would impact their role, and they needed their agreement.

One of the business units built a delivery plan knowing it would have several issues. They scheduled a working session with the business and the customer to share their new process, clearly identified the gaps in information, and asked for input on the spot, and they were able to work through a plan with the customer and the business owners. There were a lot of changes and negotiations, but they were so impressed with the work they had done together that they wanted to meet the rest of the team! Without hesitation, they all walked across the street to the engineering team's workspace and "walked the walls" together, sharing our reasoning, addressing their questions, highlighting challenges, and making suggestions along the way. After the customer left, the business expressed how much they realized that they needed to be part of this transformation team and could finally see the value of having the team connect directly to customers.

A year later this company has expanded from the first pilot to a larger scale, with many business units getting involved with the transformation, sharing with and learning from each other to find the right approach and plan. Their customers are even sending appreciations directly to their teams, managers, and leaders for all the transparency they share and how much more the company understands the customer's needs. Because the customers are so excited about this new way of working, the business leaders are starting to participate more in the transformation team and share their experiences with others that are not convinced it will work for them.

Agreements

The communication reinforcement model cannot be assumed, nor can it be eased into peripherally. The goal is an explicit, intentional change in how leaders work and think, which requires a formal transition launch to be marked with a little ceremony. This ceremony focuses on helping the executives to

1. Decide as a team, as a peer group, that they would adopt particular ways of working for this pilot
2. Agree on the timing of the new cadence of meetings to stay connected and on-track
3. Decide on a new way of tracking work and measuring progress of the pilot

This move is bold since it requires a deliberate decision to commit from executives, and it eliminates any case for them to avoid responsibility.

Through this explicit team formation activity, you can formalize the pilot team's role and get everyone on the same page with regard to behaviors, commitments, and communications. These structures will be critical when the work of leading change gets hard. When the results are not yet clear, when the old ways clash with the new, when pockets of resistance appear—that's when the team agreements will serve the leadership team. They can leverage cadences to make decisions; they can leverage communication channels to clarify; they can leverage commitments to help each other stay the course.

Sponsorship

Even with the support of a cross-departmental leadership team, the pilot also needs a single executive sponsor, the leader with just a little more skin in the game, who can keep everyone focused on results when everyday business gets in the way (Figure 3-7).

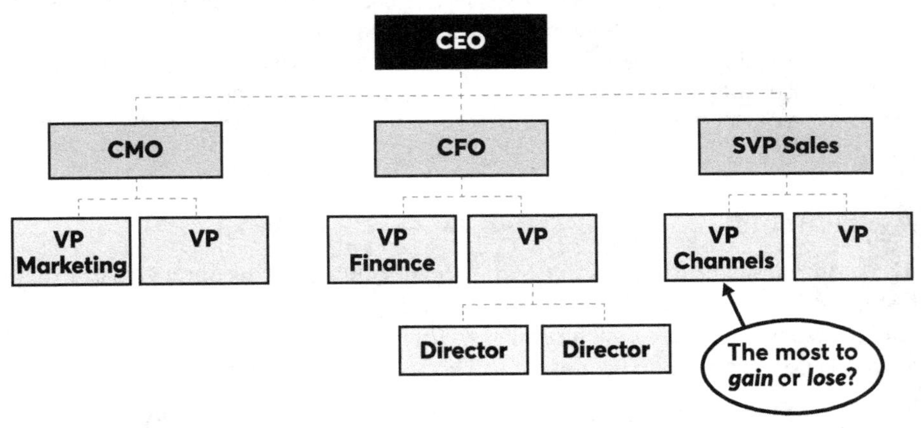

Figure 3-7. Identify your sponsor executive

Your sponsor executive must be the leader who has the most to gain or lose from the pilot's results—based on the software and the business value being delivered to the customer. With this sponsorship in place, it's now show time and you have center stage.

The sponsor has a key role in managing downward—keeping the pilot teams focused on real results, helping them focus, and in managing outward to peers, keeping the executive team apprised of progress, results, and obstacles. The sponsor keeps it all moving forward, supported by the executive team.

One important tool at the sponsor's disposal is the demonstration of working software that is helping to deliver the desired business result. This is done by all four customer-facing pillars—delivery, finance, marketing, and channels—working at the same table together with mutual responsibilities. We want to move away from the idea that finance, marketing, and channels might be considered delivery's customer. Instead, they are equal partners on this pilot.

It is through a prototype demonstration that the magic happens. This demonstration will make it clear what customer value really means. The business has an opportunity to provide valuable feedback, and the delivery, finance, marketing, and channels teams are encouraged and validated around their respective efforts and contribution. The sponsor pushes for attendance and attention at each demonstration, because they provide critical visibility and feedback. If a new way of working is going to work here, everyone needs to see progress.

With small pilot investments, companies are able to contain the risk. This is important since, as previously discussed, perceived risk can be a barrier to investing in new approaches. Pilots allow companies to experiment on a smaller scale and get to proof of concept. Successful pilots can then be scaled up with reinforcement structure (to be covered in the next chapter).

Principle Two – Decide on the MVP

In selecting what to develop and deliver during the pilot, look for the work with the "best chance of getting the most important results" for a particular line of the business. Start by finding a product or program that has significance—no one will care that you used a new way of working to deliver on that little project that serves one customer or merely checks a box. The pilot should deliver on a strategic initiative, or should deliver some value on a key project that's been stuck or is behind, or should serve a very important customer segment. It must *matter.* For your pilot, pick something important.

That said, don't pick work that is so mission-critical that the CEO is asking about it daily and there is company-wide pressure to be perfect. Perfection isn't an agile goal: approaching perfection is. When a CEO is asking about something daily, it will feel like judgment, which doesn't help with the principles of quick learning and safety. In the face of that pressure, the teams won't stick to trying agile; at the first sign of trouble they'll revert to old ways of working, either out of habit or to look for a place to put the blame, and your pilot won't prove anything.

Another factor in selecting a pilot is that it exercises the four pillars. It isn't enough to have a program that doesn't change a nondelivery process—we are not asking them to be involved as observers, so we need to have a program that will likely demand change from the nondelivery parts of the organization. Without the possibility of change in their organizations, we are less likely to have full participation. An example: One company piloted with a total re-vamp of their process and supporting technology for resolving disputes between sellers and buyers (in an online marketplace). The change involved more than just software, but also how humans would work together, and the company's very reputation.

COACHES CALL-OUT: CASE STUDY

Deema Dajani, Advisor, Transformation Consulting

Selecting the right pilot is critically important so that everyone can see the differences and impact. A few years ago, we started an agile journey with a multi-million dollar delivery group rolling up to the CIO, which was primarily focused on internal systems and applications. In the meantime, a small business group emerged around Digital Services, driving about a few million dollars in revenue. The legacy delivery organization resisted the move towards digital, viewing them as "renegades." It was a textbook risk averse reaction from the established mass majority, towards the early adopters. This is where our pilot came in—we were able to see the potential digital services and pitched an agile pilot there. Why pick that pilot? At the time it was simple—this group was starting to work on market-facing technology products. It was the only group interfacing directly with the external customers. The pilot was indeed launched, and fast forward in time—this pilot was a huge success and opened the door for more impactful agility transformation. The group itself grew over five times in size. It was the first time the C-suite took notice because of how strategic the pilot was for the company's growth strategy. The sales cycle and overall cost efficiencies improved by over an unprecedented 70%. One leader memorialized it with this quote "It was not about becoming agile, it was about getting value out to our external customers so they could fall in love with our brand and garner growth through long-time brand loyalty." Two lessons from this story: Embrace change, and pick the "right" pilot to create an organizational tipping point.

Having chosen the right program, the pilot team should help guide the teams to focus on the right subset of work. There's never a shortage of features that the team will want to deliver. But the opportunity here is to make a material impact in a very short period of time—two months is ideal. Therefore, identify the one thing you can realistically deliver that has the most customer value in that timeframe. A useful concept here is MVP.

Minimalism is your friend. What's the least you can deliver that will still delight your target customer and provide real value—the kind of value that keeps them coming back for more? The idea is to drive conversations about what we know—sure value—and what we don't know, and therefore what we could build that would help us learn what to build next.

Viability is essential. What you deliver can't be so small that no one takes notice. What you deliver must solve a real problem, and there might be some tradeoffs that you'll need to make. Look for something that is of enough value that it's worth the investment and the risk.

Note that the MVP also sets up a perfect pivot point. The MVP is a bet about what the customer will value and what will deliver impact. If you bet wrong, or if the feedback shows you were slightly off, you haven't invested too much. Our human tendency to overvalue "sunk cost" can be overridden, and we can pivot. The MVP sets up learning and homing in on results.

Getting to the MVP is often the art of narrowing focus and getting laser-focused on value (see the next section). This technique will continue to serve us as we scale the new way of working, but it is absolutely essential for the pilot. Without it, you'll have a group of people thrilled and proud of their work, but the rest of the organization will be underwhelmed with the outcome. With the MVP, you have a chance to really inspire.

Having chosen the MVP, schedule the showcase two months out. Invite everyone, casting as a wide a net as possible, and remembering all four departments—delivery, finance, marketing, and channels. Now leverage the focus to deliver.

ROI Decision Map to Determine MVP

The return on investment (ROI) map is a tool that can be used to help guide conversations and establish what feature functionality is more valuable than others, and what capabilities would potentially yield higher ROI. The technique to using this tool centers around the business representing the voice of the customer and being able to prioritize product functionality on the basis of the value to the customer. Then, others represent the technical voice and prioritize based on technical complexity (which includes time and cost).

The results of these discussions will provide a visual representation of decision making, called the relative ROI decision map. This map will help to drive further conversations about the MVP to understand what functionality has the highest business value and should be delivered first.

For proper decision making regarding what product functionality to build, and in what sequence, establish this relative ROI decision map. This is made up of considerations from two key criteria:

1. The ability to generate revenue or profit from this product decision. Taking the concepts of the negative impact of the cost of delay, as Don Reinertsen[1] defines it: "What is the business value that we would be deferring by not implementing this product decision." In other words, it is the opportunity cost of not doing something. In this relative ROI decision map, we turn this negative statement into a positive aspiration while leveraging the same concept by asking "How much can the product generate revenue?"
2. The ability for delivery to actually deliver this. What is the pilot timeframe to deliver this?

The approach shows that the business has their say on the value of each item they want built based on the business value, as established according to the voice of the customer. Depending on the product being built, this can also include the voice of the business experts in this product domain for any new feature and capability that they are trying to drive.

By weighing these answers against potential decisions, you will build a visual map of where the priorities should be (see Figure 3-8). Decisions that have a high potential revenue (or as Donald Reinertsen would describe it, the high cost of delay for failing to act) and a fairly short time to market should be worked on immediately. On the opposite end, decisions that have a low revenue potential or low cost of delay, and that have a long time to market, should be deferred.

[1]Donald G. Reinertsen, *The Principles of Product Development Flow: Second Generation Lean Product Development* (Celeritas, 2009).

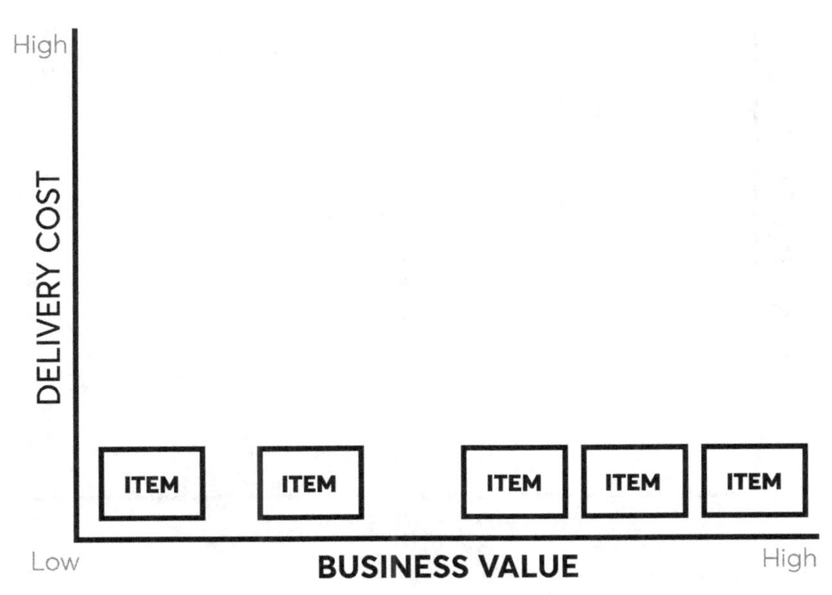

Figure 3-8. Visual map of priorities

This is shown on the horizontal axis in the diagram for business value scale from low to high. The technique here is for the business owners to have their say by determining business value horizontally from low to high. The business owners cannot move items vertically.

Next, this shows that the technical folks (including delivery, at least in some companies) have their say on how much time/cost it will take to produce the feature functionality that the business/customer wants. This is shown on the vertical axis in Figure 3-9 for delivery cost. They cannot move items horizontally.

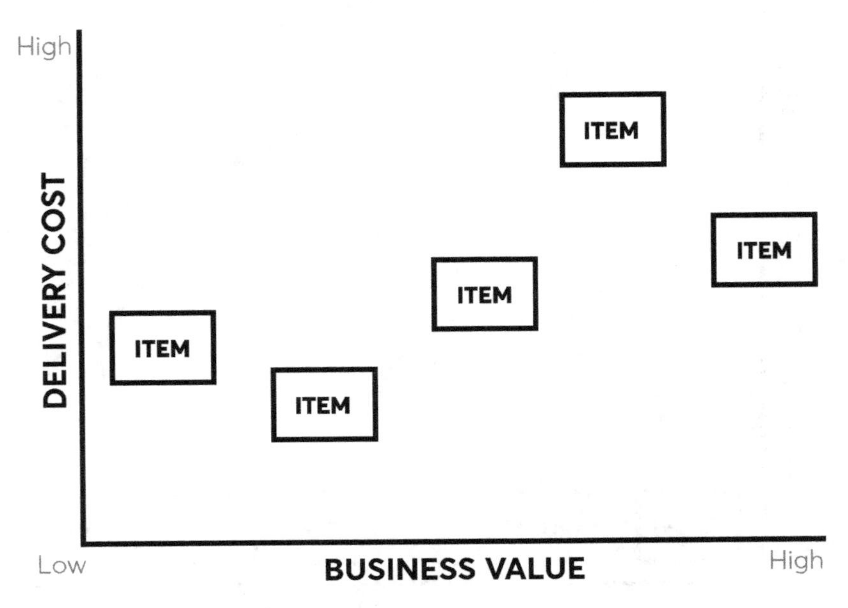

Figure 3-9. Rank priority business value

Now that the business and technical owners have had their inputs, the group will find some natural breaks; you'll see where you can draw arcs to group initiatives together in Figure 3-10. The lower-right-most group has the highest ROI, and as we work our way out on the map, ROI decreases. Note that items with low value and low development costs (in the lower left corner of Figure 3-9) usually end up being deleted or archived as there is newer, more interesting, and higher-value work to do.

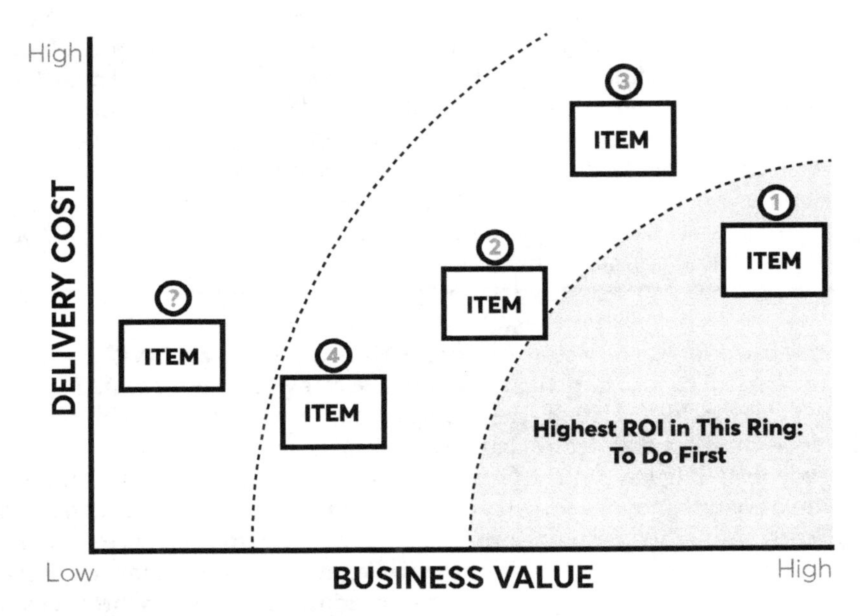

Figure 3-10. Reality check on priorities

In addition, when considering items in the upper right-hand corner that are high value and high cost, the group must be challenged to think about how it can be broken down into smaller deliverables. Often there's a valuable less-expensive gem hidden in the bigger work item that can be discovered by using the relative ROI model. Once you have broken down the large work item into smaller pieces, return to the ROI model and this next round of discussions should be about those smaller and valuable pieces of work.

The results of a completed pass of what the map shows is a great first step to having a prioritized grouping of functionality to start creating a product roadmap. When maintaining and iterating on this map over time, this tool and technique can help any pivot or trade-off discussions on product functionality that may be necessary when feedback is obtained from the business and customers.

ENGAGE PRODUCT STRATEGIST DURING THE PILOT

As we've already touched on, choosing the right pilot is critical. And breaking down that pilot into the smallest solution that provides customer value – the MVP – is its own challenge. Product strategy will most often facilitate the activities involved in both, including the work that goes into the relative ROI decision map. Because product strategy is the customer proxy, they have to synthesize the customer requests and the features brought forward by the four pillars. They have the difficult responsibility of making long-term strategic decisions that will drive new customers and retain the existing customer base, rather than looking at short-term gains that don't improve retention rates or gain market share.

So how can long-term vision coexist with the pilots and MVP? It's simple but very hard to do, and many product strategists overestimate what is needed for the MVP. Use all the inputs available from your customers, the ROI decision map, data from channels and sales, and last but not least, your *best judgment* (Figure 3-11). The point of a good product strategy is to provide a roadmap of the long term. Pilots and MVPs, on the other hand, are used to testing your assumptions about one of those long-term, complex, and expensive roadmap items. Start small, get customer feedback, iterate, and pivot if you need to. A good product roadmap can absorb the necessary changes in direction, and the MVP will, early in the process, help drive a future holistic solution on that roadmap.

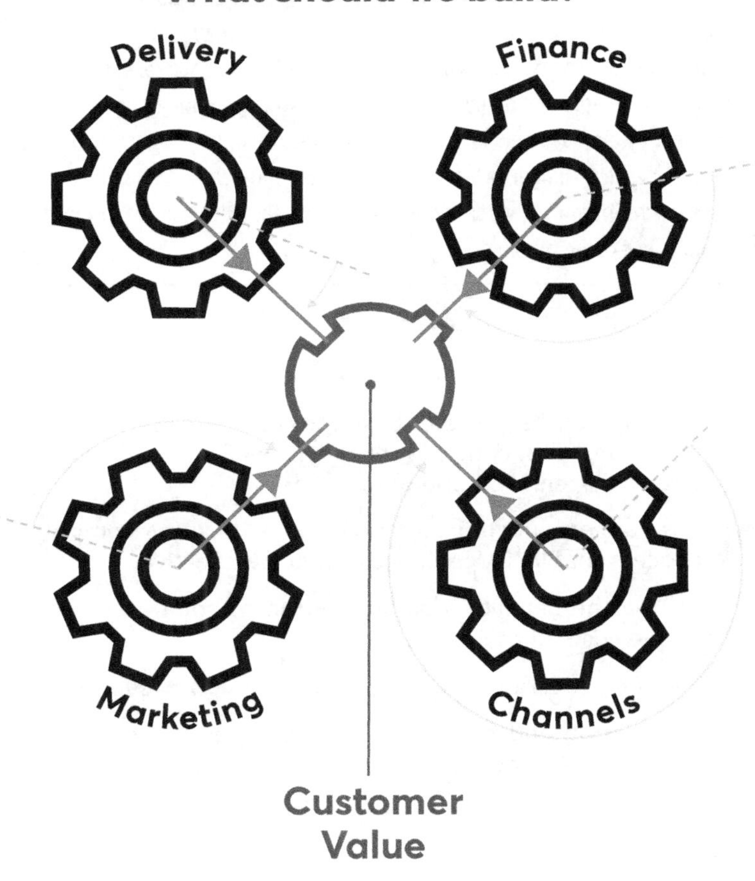

Figure 3-11. Product strategy

Set the Pilot Demo

Set up a conference room and invite everyone so there is standing room only for the demo in those two months. The demo will show the actual software with a few features working, and tell the story of how the actual customer would use it. This minimum amount of functionality to tell the story is otherwise known as an MVP demo. The demo will show key functionality that was tangible from a customer's perspective and in language that the executives understand.

Before you have a shine-the-light conversation, it's important to verify that the executives have the awareness of the company problem and knowledge that changing their operating model will be a part of the solution. Therefore, each executive will be open to change to deal with the company problem. This open-to-change mindset can be based on their own understanding or can have been revealed from previous conversations with you: the executive has awareness about the company problem of digital disruptors and demanding customers, which you might have sounded the alarm about, or maybe they already knew it. This is the knowledge and motivation that executives need to change their operating model, which they might have known already or might have gotten from your look-in-the-mirror conversation that had them face the hard truth that the executives themselves are a part of the problem.

Let's Review

- Tactfully challenge the business to find better ways of solving the company problem, while introducing a customer-centered operating model
- Lead the executives in creating a vision of a new company process that includes the customer perspective
- Sell on the benefits of an agile business solution related to the goals of the executive
- Create a coalition to create a shared vision of success
- Take action and conduct a live pilot demonstration to prove success and establish a tipping point toward change

CHAPTER

4

Agile Business Realization

Your successful pilot, during the shine-the-light stage, demonstrated an agile business in action, with a customer-centered operating model that delivers customer value quickly and enables the company to compete with agile digital disruptors. The executives have seen the light: they witnessed the ability to solve the company problem on a small scale by delivering customer value quickly. You reached a tipping point with this successful prototype and there is general optimism, with everyone seeing the benefits of an agile business, a customer-centered operating model, in action.

Even when you have everyone agreeing on the merits of a customer-centered operating model from a successful pilot, which is to deliver faster and iterate toward delivery of customer value, there is still a challenge when you want to grow this company-wide on a larger scale and have the company achieve a sustainable business velocity in this. From your previous small pilot, the challenge now is how to earn larger executive commitment and establish a customer-centered operating model as the new standard in your organization. In order to capitalize on this tipping point and successfully create large-scale adoption in the organization, you will need to introduce a reinforcement model that achieves agile business realization.

Agile business realization is not achieved by simply delivering products that the customer values. That is only the beginning. Now, you must go the next step in the cycle, beyond product delivery, and produce actual business results from the delivery. Realization is achieved when the customer uses the

J. Orvos, *Achieving Business Agility*, https://doi.org/10.1007/978-1-4842-3855-4_4

product, which accomplishes the company's desired business results. Agile business realization implements the new customer-centered operating model on a large scale so that the organization can create and maintain sustainable business velocity and achieve measurable business results. To accomplish this final outcome, you will need a reinforcement strategy that holds executives accountable to each other to deliver customer value that leads to achieving business results, that is, agile business realization.

What Are They Thinking?

While there may be increasing openness to the need to adapt agile, recent studies have identified an alarming lack of confidence about making change happen in a timely way. For instance, in a 2016 study by Scott D. Anthony, S. Patrick Viguerie, and Andrew Waldeck,[1] 66% of participants agreed or strongly agreed with the need to change their core offerings or business model in response to rapidly changing markets and disruption. At the same time, however, only 15% said they were "very confident" that they could achieve a transformation over the next five to ten years.

What accounts for this confidence issue? One major factor is that organizations default to flawed operating procedures that benefit their own departments at the expense of the growth strategy of the company. In response to a question about the "organization's biggest obstacle to transform in response to market change and disruption," 40% of survey respondents blamed "day-to-day decisions that undermine our stated strategy to change."

The executives from delivery, finance, marketing, and channels are happy the pilot was a success, yet still skeptical that they can expand this on a large scale in the organization. This is because they know there are many more challenges they will encounter and are unsure they can overcome them. When these executives consider the daunting effort to expand this pilot on a larger scale in the company, they will be aware of the risks. Change efforts are hard. The executives might rather settle back to the status quo. They're now feeling the pressure to make the right choice. They may be thinking, "Do I really need to make this change since I've been successful up to this point? What if it's the wrong decision and this ultimately fails?"

[1] *Innosight Corporate Longevity: Turbulence Ahead for Large Organizations*, Spring 2016.

Many executives will sit on the fence when it comes to making significant decisions that can influence the way others in the organization perceive them. Executives may not say "no" to you directly, but they might rather hold back any announcement of support, so they can hedge their bets. Until they see enough support with their executive peers, the ability to change an organization will hang in the balance.

The executive optimism from the successful pilot dissipates and is overcome by anxiety when considering the risk to expand this effort on a larger scale. Your executives realize they need to make a decision either to formally support your cause for an agile business change or to step away. During this period of high anxiety, the executives are looking for a plan that provides assurance that the small pilot success could be implemented on a larger scale. The assurance plan will give them confidence it can work as long as it comes from a trusted source—their peers.

Industrializing Long-Term Change

To manage beyond the tipping point and establish long-term change, your reinforcement strategy should include the following: ***creating an agile business realization plan*** and ***changing the organization culture***.

Creating an Agile Business Realization Plan

To sustain change, create a reinforcement plan that builds upon the principles you learned during your pilot: build a pilot team and decide on an MVP.

The goal of this plan is reinforcing the skills learned from the pilot to be applied to a large-scale adoption across the four business pillars. As stated in earlier chapters, delivery, finance, marketing, and channels need to be able to work in unison for the organization to steer and respond to changes quickly.

At this point, expand upon these principles on a larger company-wide scale. Success is based on creating peer accountability to deliver business results. To achieve agile business realization, your plan must

1. Establish a leadership coalition (reinforced and expanded from the pilot team)
2. Drive business results (reinforced and expanded from the pilot MVP)

Let's examine each of these in a bit more detail:

Establish a Leadership Coalition

During the pilot, you acquired the ability to build a small cross-functional leadership team to sponsor this effort in a short term with a specific goal to demonstrate the customer value. To build on this capacity and expand to a larger scale, establish a peer leadership coalition among leaders from delivery, finance, marketing, and channels that interacts and communicates under a new social contract to work as a community. A leadership coalition is responsible for the large-scale success of this customer-centered operating model change effort, and that becomes the force for the large-scale change and agile business realization. The leadership coalition will communicate the reinforcement plan (to be outlined in this chapter), to empower peers to help make the change grow on a large scale.

This peer leadership coalition allows this group to investigate a product opportunity from many perspectives, which allows them to make informed decisions—not just from siloed perspectives, but from the system as a whole. These separate customer-facing pillars will now think of themselves ultimately as one team with the common goal of delivering customer value and will gain mutual understanding and respect.

This leadership coalition is based on creating peer accountability to deliver business results. The positive peer pressure can be the lever for accountability. This approach enlists the leaders to take ownership and make a plan together to build their confidence for large-scale success.

Coalition Guidelines

The coalition guidelines are as follows:

1. The shared coalition motivation will find a common ground and alignment around the common goal of delivering customer value. Therefore, the team will come together and form a bond with empathy with the others. Coalition teamwork also allows the group to investigate a problem or opportunity from many perspectives through brainstorming, which allows a team to incorporate different perspectives and ideas.

2. The coalition will communicate all successes and failures regularly and consistently to the organization. This transparency is important so that the executive peers know the business results of their decisions. This will greatly help the large-scale expansion stick and become the "new normal" in the organization.

3. The coalition understands that only together as a team can they accomplish the goal of delivering customer value. The respective executives, as a coalition, can accomplish large company goals that would normally be too complex or large for an individual department to do alone. Alone, each individual business area may fail in its efforts. As a four-part community, this one cohesive unit increases its chances of success.
4. The members of the coalition are open to learning from each other and welcome different insights to make compromises that help accomplish the common goal of customer value. Together, they will see exponential improvement and truly seek to help each other while their own line of business succeeds. They will become focused on the team for mutual success, instead of myopic self-interest.
5. The coalition as a whole will be recognized for its success, meanwhile the coalition members will support each other when mistakes are made. Teamwork with the customer-facing pillars involves the interaction of individuals for a common purpose where the interests of the individuals are not as important as group unity and the effectiveness of the group to succeed as a whole. This coalition model creates the shared, common goal of funding, developing, promoting, and selling the high-value software and will help the teams' parts thrive together.

Leadership Coalition: Four Pillars

The teaming of the customer-facing pillars is an all-or-nothing ordeal, meaning each pillar executive needs to be onboard with the change, because they all rely on each other. If all four customer-facing pillars do not come to the table, the entire effort will fail or wallow in mediocrity with unpredictable product hits and misses. The customer-facing pillars are misaligned when one of the business executives, whether from delivery, finance, marketing, or channels, does not participate. Think of these four pillars as four legs to a stool. If one leg (pillar) is missing, the stool will fall (Figure 4-1).

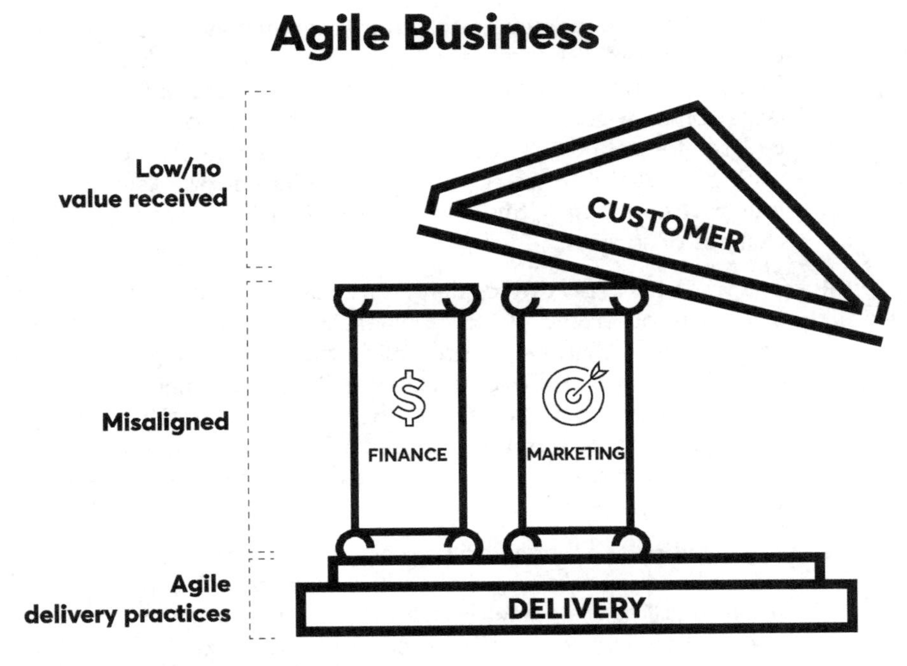

Figure 4-1. Non-agile business

The coalition is misaligned when one customer-facing pillar accomplishes its task, but the rest of the company cannot build or support it; then the company needs to wait for the rest of the customer-facing pillars to catch up or realign. A new product can only deliver business results when it reaches the customer and it's used. Therefore, the company needs to get software innovations into customers' hands to monetize their work. For example, finance funds software that delivery builds—say, a new mobile platform or a new set of features that might delight users—but the channel is not ready to sell or support it, and marketing isn't ready to promote it. Although delivery can develop new and innovative software, the company will not be able to launch it and the customer will not receive any value from all this work. In this situation, while your company will be waiting for months to get the other business executives in line, your competitors can quickly make a move and launch their alternative product and take market share away. Your company will need to make changes in order to be able to have speed, otherwise launching new products and services slows down and customer value is not realized—remember, your company is as fast as your slowest link. Now, imagine how finance, marketing, and channels feel in this situation. They will be overwhelmed, which is not sustainable to build a coalition for the long-term success of any organization. For those that remember the *I Love Lucy* chocolate episode, there is a scene where Lucy and Ethel have the job of wrapping chocolates coming off a machine conveyor belt. When the chocolates come out too fast, chaos ensues in this iconic comedy scene.

In this leadership peer coalition, all four customer-facing pillars need to work in unison to achieve and sustain the high maximum velocity potential of an agile business. This means that delivery can build it, finance can fund it, marketing can promote it, and channels can sell it—simultaneously. When aligned, the leadership coalition can deliver business results and pivot across delivery, finance, marketing, and channels. They can change as needed because they are aligned, dependent on each other to deliver business results: delivery and finance rely on channels and marketing to understand the real-time customer perspectives to develop and properly price the product; finance needs delivery to make sure the software they fund is developed; marketing and channels need delivery to release what they promise and keep them informed of any changes along the way. Everyone needs everyone to be a part of the commitment to deliver business results.

Drive Business Results

With the coalition guidelines in place, there are the two important meeting cadences that will enable the whole organization to deliver business results. In quarterly business steering meetings the extended leadership group collaborates to agree on corporate strategy as well as to commit to enterprise improvement initiatives (that is, to work both *in* the business and *on* the business). In short cadenced pivot meetings, leaders hold each other accountable for the strategy, while reviewing feedback and making adjustments as needed. As a result of these two meetings, the company will be able to quickly respond to changing tastes of customer value in the marketplace, as well as new information about what is and is not working within the organization. While annual planning helps set enterprise targets and guidelines, these meetings make it easier to adapt to the ever-shifting competitive landscape. Now, with these two meetings, the company will be in a position to decide to pivot or stay the course because of their new ability to quickly respond and deliver customer value, which is the source of a competitive advantage.

In order to set this new way of working, first set a kickoff meeting to establish that the customer is at the center of all decisions and that the common goal is to deliver customer-valued products that achieve the desired business results. In this kickoff, explain that the old norms no longer apply. Root the conversation in the results and findings from the pilot. Agile is about reality-driven planning, so explain agile ground rules and expectations: open collaboration, speed to the customer value goal, welcoming the unexpected, basing decisions on feedback from experiments instead of politics, and personal accountability.

Quarterly Business Steering Meetings

Quarterly meetings are the rhythm that brings the peer leadership coalition together to focus on building and refining the business strategy to deliver and achieve value realization. This is by no means another stodgy executive meeting where the results are known and dictated. Instead, this meeting provides a forum for leadership to collaborate across their functional lines, and for the decisions to emerge during the session. If and when a pivot is needed, this forum allows the whole company with all its departments to pivot together.

Depending on the priority in a given quarter, these sessions can have two areas of focus:

1. Delivering customer value. Working *in* the business. This is where decisions are made about such things as: building a new trial product; taking a new position in the market to serve new customers via thought leadership; or launching a new sales campaign to target a new market segment.
2. Improving the customer-centered operating model (to deliver the customer value). Working *on* the business. This is where decisions are being made to tackle broken or suboptimal parts of the people/process/technology that runs the business. You might, for example, charter a cross-departmental initiative to fix the idea-to-delivery process, or to break down barriers through an initiative to better connect specific teams from delivery, finance, marketing, and channels.

Focus on Delivering Customer Value

Delivery, finance, marketing, and channels—and potentially other departments—must ALL come together on a quarterly basis to build consensus and align to the customer value being delivered. Delivery must improve to be able to respond and satisfy these changing customer desires. Delivery groups will align closer to the customer journey if they can count on intelligence from marketing and channels. In response to what they learn, delivery and finance can reorganize around customer value. By the same token, this focus allows for effective pivots because the pivot is a response to what they learn. Hence, the company will respond faster with customer-valued products.

The intent is for everyone to discuss this strategic plan as it relates to their respective department's needs, identify conflicts or dependencies, and discover how each pillar may affect others. All parties can ask questions, identify issues, and otherwise ensure a single vision in moving the plan forward. When issues are identified, each department executive and/or leader has open communication to determine how best to mitigate any problems or dependencies.

The quarterly business steering meeting will help all four pillars evaluate their current strategies, recent learnings, and competitive and market intel to decide whether to stay the course or reevaluate. By combining rear-view and forward-looking perspectives, they can create realistic plans with confidence around the ability to deliver each quarter. These quarterly strategic meetings provide a chance to look backward, conduct a retrospective on what the organization has learned, and reflect on what has been delivered into the market. Every department must review its own and the other departments' performance metrics to understand both outcomes and learnings. The meeting itself (or the preparation for it) is a chance to query each other—not to so much to challenge, as to understand and strive for a truly holistic view of the business and its performance, including performance in learning. The lookback then feeds into the forward planning.

During these quarterly strategy sessions, all pillars pull together to figure out how they might collectively meet the business objectives by delivering software and surrounding products and services that provide value to the customer. Planning on building the right products with these different departments is complicated, with competing priorities, constraints, and dependencies. Therefore, conversations that focus on making decisions about plausible plans that are agreed to by all are the most important to provide customer value. These meetings result in a creating a shared vision of the product to develop.

COACH'S CALL-OUT: KNOW YOUR BUSINESS

Deema Dajani, Advisor, Transformation Consulting

To know your business is "the prerequisite" to business agility and the quarterly steering meetings.

Customers have asked me how these quarterly steering meetings have become so effective. While it is true that there is a secret sauce in the facilitation of such events, the truth is, it is only an event. You need a solid business foundation in your day-to-day operations; otherwise you may be pivoting without purpose, which would be disruptive to the organization. The link is understanding where you have gaps in your core capabilities (e.g., pricing capability), and strengthen them via initiatives discussed in the quarterly steering. You also need the environment and capabilities in place that enable "sensing" changing needs when or before they happen! This includes sensing changes from customers, regulators, and competitors. This ability to sense and respond to market changes as a matter of everyday business is how we define business agility.

How do you know if the pivot is the right answer? You need to understand your business fundamentals and you need to understand your customer. These are your prerequisites.

In preparation for each quarterly business steering session, the driving question that must be answered is: "Given the business strategies for the year, the company's current performance against the targets or metrics, and the peer leadership coalition's understanding of organizational health, what should the focus of the leadership coalition be in order to improve our likelihood of short- and long-term performance?"

These strategic quarterly steering meetings will coordinate across all pillars to collaborate and remove risks and obstacles to achieving their business objective. They will address the following questions, which run vertically from the top vision to the bottom. There are "vertical questions," which start discussions from the top executive level perspective and drive down to the predicted realities implementing the strategy. By combining all levels of views during these steering meetings (top executive to bottom implementation), the group will come up with the agreed quarterly product and operations plan.

Overall Company Strategy

Start from the top and remind the team about the overall business strategy and corporate vision. This simply comes from the overall vision of the company has set, usually outlined by the CEO over the next few years, as a reminder to all the executives that this larger business strategy is in place and the goal is to focus on this. What was the overall product vision to implement the company strategy?

Agree How to Measure Customer Value

Determining what metrics to measure can be daunting. There are numerous indicators that the peer leadership coalition can measure, and sometimes the thinking is the more metrics, the better. The inverse is true; actually, a few very targeted and specific metrics are the most effective way to measure customer value. What are these metrics and how does the peer leadership coalition properly determine them? Implement the following approach to define metrics:

- Business goal

  The peer leadership coalition starts with measurable business goals. Some examples are increasing customer satisfaction, penetrating new market segment, building competitive advantages to competitor X, maintaining compliance, or being best in class.

  Metrics are based on the insights learned from decisions to accomplish the specific set of business outcomes, not delivery output goals. Make decisions based on the desired business outcomes, and look for insights that will help define your measurements.

- Product decisions based on the business goal

  Who is the target customer? What are the customer segments? What is the problem to be solved? How can value acceptable to the customer be delivered?

- Insights learned from these decisions

  Do we all know what was delivered? Was it as planned, or did we pivot to deliver something else? And for each result: Why? Most importantly, what did we learn? Did we get customer feedback on our capabilities on the product? Did Wall Street analysts give us feedback about how they see our business? What did we learn about how our message resonates, or whether our offerings are competitive? In a nutshell: What new information do we have to guide our next decisions?

- Metrics to illustrate the insights

  Based on learnings that are clearly illustrated from the metrics, the team can make informed decisions on any proposed changes this quarter.

What did we learn from last quarter about strategy results and challenges?

In order to have the "realization" of an agile business, the learnings must be focused on the business response to your decisions, not simply that products were delivered. Realization only comes reviewing business results based on product use. So the focus on learning is on the business results.

Since each pillar spends its time heads-down executing their part of the strategy, this is a chance to pause and reflect upon the big picture. This is the opportunity to have a retrospective to develop a shared sense of the product strategy by focusing the executives on the results of current overall strategy and developing a shared understanding of the effectiveness. It then looks at the competitive landscape, especially of changes since the last quarterly steering meeting.

What have we learned about what we have delivered or about our business goal? Do we pivot or preserve? The peer leadership coalition must be on the same page. Therefore, institute the business metrics that have been established during this quarterly meeting (earlier in "Focus on Delivering Customer Value" in this chapter).

What is the product go-to-market for this (next) quarter?

In an agile business, there is no longer such a process or outlook as delivery simply building products that are dreamed up by a product owner. This has now matured into a tangible product go-to-market plan and includes the alignment of all four pillars.

Based on what was learned, what ideas are there to capitalize and deliver value to the customer this quarter?

What is the agreed product vision to accomplish this strategy? That is, given the inputs described in the preceding, what adjustments do we want to make to our strategy? The focus here is on whether to stay the course, to adjust, or to pivot entirely. Rearticulate our hypotheses.

What will the delivered business value to the customer be this quarter? Given those hypotheses, what do we actually plan to deliver, and how do we expect customers to respond? Are we ready to support the response?

What has been decided this quarter:

- What is the unique value proposition of the product vision?
- What problem is the product solving?
- What are the product advances or advantages?
- What are the channels and path to customers?
- What are the expected revenue streams and profits?
- What are the metrics for success?

What will the company deliver this (next) quarter?

It's not just about building products from delivery anymore, but rather delivering them to the customers to create business results. Therefore, the leadership coalition—with delivery, finance, marketing, and channels—decides, as a team, what they can realistically deliver this quarter.

How much work can a "product system" (which includes all pillars in a room) take on this quarter? What's standing in the way of everyone committing to this quarterly plan? What are the dependencies and risks with other areas of the business? Can we mitigate them now, or shall we just accept and track them?

How are the possibilities prioritized? Use the *relevant ROI decision-making* model you learned during the pilot for this decision-making progress. Setting stack-ranked priorities prepares us—and people down the chain of command—to make trade-off decisions later. With a clear understanding of priorities, more people are empowered to quickly make good small-pivot decisions.

What has been decided this quarter:

- How to deliver value acceptable to the customer?
- How are the possibilities prioritized based on data?
- How much work can we take on this quarter?
- What is the cost structure to build this?
- What are dependencies/risks within the business?

How can we better sense and respond this (next) quarter?

How accurate was your sense and how effective was your response last quarter? How can you improve in sensing opportunities or threats and responding by delivering customer value? What operational model improvements do we need to continue to accomplish these improvements? This meeting is not just about what product to build. It must also be about the continuous improvements of the customer-centered operating model and identifying the organizational improvements that will drive toward continued success (Figure 4-2).

Quarterly Steering

The Business Strategy?

What did we learn from last quarter?
What is the product go-to-market this quarter?
What will the company deliver this quarter?
How can we better sense and respond this quarter?

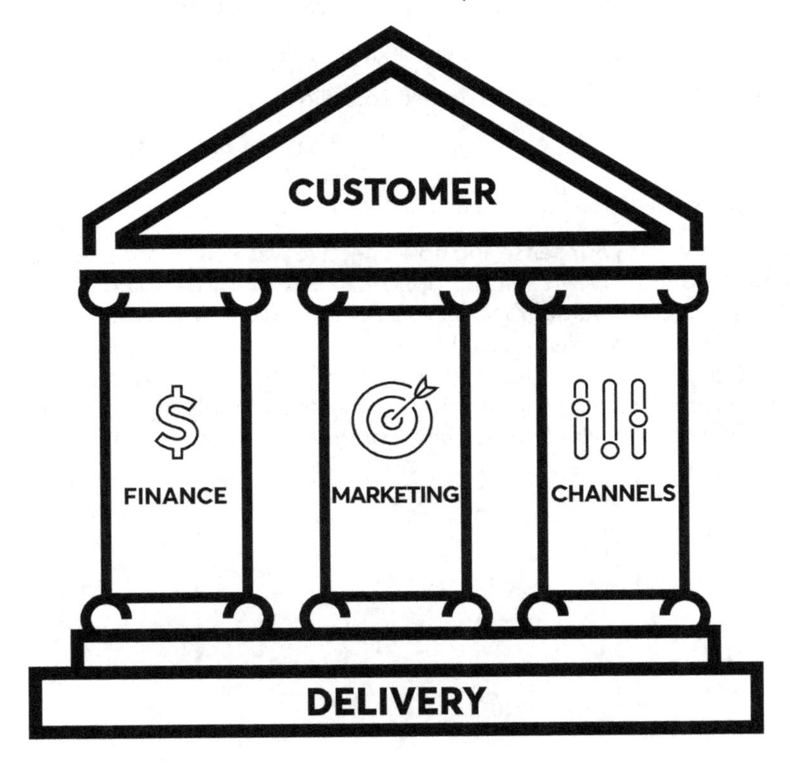

Figure 4-2. Quarterly steering

Focus on Improving the Customer-Centered Operating Model

In the quarterly business steering meetings, the team can also work on improving the business—the organizational improvement initiatives designed to help the company succeed—to collaborate better, to streamline systems, and to remove bottlenecks.

Now that we have gleaned so many learnings about delivering customer value, we will be held accountable for realizing business results. During the quarterly and cadence meetings, we want to focus on reinforcing the customer-centered operating model best practices established during the pilot. In addition to working to determine what products and features to build to serve customers best and deliver value, the leadership group also needs to work on improving the customer-centered operating model. What improvement initiatives should the organization undertake in order to improve its ability to deliver customer value and support the business?

The initiatives this group charters and monitors would be the ones that cross departments. When two or more departments must collaborate to fix a process, or improve a workflow, or solve a problem, then that work should be elevated to be visible to the whole value stream leadership. This work takes precious resources and so shouldn't be allowed to fly under the radar. If it's a priority, it must be discussed and agreed to, and then tracked.

This continuous improvement requires yet another mindset shift from leaders. Where they used to hold people accountable for failures in the system, they now must ask simply "What did you learn?" and "What improvements have you found?" The only failure is failure to learn.

These strategic meetings will coordinate across all pillars to collaborate and remove risks and obstacles.

Alignment of Goals

Of course, every department is doing its own work to serve the business. We don't need this larger group to micromanage departmental efforts. However, we do want the group to have a whole-system view of what departmental work is happening. If pillars are misaligned in their priorities, this is the meeting that should make that visible and invite a conversation.

In general, the quarterly steering meeting is very much about alignment. When organizations work inside silos, it is so easy to get out of sync, to be working on counterproductive priorities—not through maliciousness, but just through a lack of visibility and communication.

All four customer-facing pillars, delivery, finance, marketing, and channels, must work together as a coordinated team and keep the common quarterly goal in mind and their activities transparent. If they are uncoordinated, and one customer-facing pillar goes rogue and races ahead with non-quarterly goal activities, it is misaligned and may damage the trust from their leadership team peers. For example:

1. Finance, marketing, or channels charge forward to support anticipated products that they believe will be created by delivery, but they later find out that never happens.
2. Delivery spends time frantically building solutions but finance isn't ready to fund, or marketing isn't ready to promote, or channels aren't ready to sell and support.

In the preparation for quarterly steering and in the meeting itself, everyone has a chance to reflect on their work, to see the whole of the work in the system, and to verify alignment to strategic goals. Strategic goals are restated and clarified, giving everyone a chance to reset onto the same page and then reassess the business strategy. When work appears out of alignment for delivering on that strategy, this meeting is the chance to discuss what adjustments are needed.

LEARN TO ADAPT

Many attempts at operating model improvement have failed over time. To ensure success, the leadership coalition needs to understand that its customer-centered operating model will develop and improve over time, and apply that understanding to learning to adopt these improvements.

Given that the leadership coalition's product go-to-market activities will glean results in short time frames and small batches, they have the ideal setting to learn and adapt. Since delivery has already been practicing agile, this is nothing new to them. This practice of inspecting and adapting has worked in smaller batches for some time. Now, we can extrapolate these same principles to the other pillars: finance, marketing, and channels.

The product go-to-market plan provides the overall guidance. But, it is even more important, just like a boxer does in a ring, to be able to adapt, as we are in the heat of the moment, based on changing conditions. Therefore, regular review and retrospectives are important. Step back and "look in the mirror" to reexamine how your operating model is solving the company problem and delivering customer value. Reexamine your assumptions based on real information and adjust accordingly.

These are the learnings we want to make sure that we keep sticky in the organization going forward so that they become anchors, and then habits for how you work.

Continue Using Relative ROI Decision Making

During the pilot, you learned how to make decisions using relative ROI evaluation. Now build upon this during the collaborative exercise used every quarterly planning cycle to focus everyone on the highest-ROI efforts to move the needle for your business. Focus begins by emphasizing the work in the lower-right arc—those are the high-value, low-effort initiatives that bring value quickly.

As the organization matures its planning practices, it learns that some of the lower-ROI items can be adjusted to make the cut. The conversation goes something like this:

> *Business Leader A: That one item has so much potential and value. I can't believe it's not making the cut! Why is it so large?*
>
> *Technology Leader B: Well, it requires [insert huge tech effort here].*
>
> *Biz Leader A: Is there any way to implement cheaper?*
>
> *Tech Leader B: Well, if you removed [challenging requirement], then the effort would come way down.*
>
> *Biz Leader A: Great! [Challenging requirement] isn't the important part, anyway. Let's descope and fit that initiative in!*

The organization that gets good at that kind of collaborative reframing and rescoping will get good at focusing everyone's effort on the best work. In addition, note that the items in the lower right area with low business value and low development cost usually end up falling off the list. It's okay to delete or archive items off the list. At the next quarterly planning, there will be new and more interesting work that is of higher importance to the customer.

COACH'S CALL-OUT: INCREMENTAL BUSINESS RESULTS

Marie Kalliney, Agile Managing Consultant

I was once asked to facilitate a business agility pilot with a financial services company. As with all organizations, this department was under water with their list of products and features. One of the many challenges they faced was that they were unable to provide customers the ability to make instant virtual payments. While they could print personal checks on demand and send them through the mail, major selling and buying platforms and nearly every major retailer in the US processed only online payments.

They were working on items attached to a long term multi-year plan, rather than running a business that could cope with market demands. Digital disruptors such as PayPal and Amazon were coming for their business and online payment platforms were easily winning market share. Honestly, I wasn't a banking expert at the time but I could tell a lot of their initiatives were already outdated and their competitors had already solved those problems years earlier.

On the surface, the challenge was extremely complex for solely the delivery organization. But when we took a closer look at how they were running their business, the finance, marketing, and channels divisions were no more ready for a new plan than the delivery division. They invariably chose the long term plan over shorter term wins. Most executives appeared so personally attached to their plans that pivoting seemed to be regarded as failure than instead of the success that it eventually was. This cultural hurdle was one of the hardest to overcome but I sought out change agents and other respected executives who helped their peers overcome their fears and firmly held beliefs in long term plans. It took time, perseverance, and carefully applying our strategies and messaging at opportune times.

We needed to find a way to deliver incremental features that would eventually lead to a fully robust solution that could compete. We had at least one enormous advantage over the disruptors—we were the backbone that funded the PayPal and Amazon accounts in the first place. In order for customers to use those other platforms, they had to have a checking account or a credit card that we provided, but we knew we couldn't provide online payments all at once. It would have taken us years to build a formidable product, and at the same time, we couldn't ignore the problem any longer. We forced ourselves to think of smaller batches of functionality that would be steps in the right direction, and with the right frame of mind, it did not take much time for ideas to start coming in from all directions. Channels and marketing were finally being respected as the voice of the customer, and delivery, although already agile, had to figure out how to deliver smaller batches while maintaining the integrity of the larger system. Even finance, which had been asking departments for yearly plans for dozens of years, eventually began embracing funding long lived agile teams instead of annual plans.

We scrapped the five-year plan to deliver a competing product all at once, and we started by offering customers the ability to transfer money from one account to another, so long as both accounts were held by the bank. That took us ten months, but it helped lay the new operational foundation across the entire organization. We used that pilot's success to build momentum and six months later, we successfully delivered the ability for customers to transfer money between different banks. That's how it all began—ideating and creating incremental progress for a very large and complex problem, both for the business and for technology. While they missed the mark with online payments, they learned some invaluable lessons about pivoting and embracing an emergent marketplace. The very same bank that still fails to compete with PayPal and Venmo is the market leader and is paving the way for cryptocurrency.

Improving the MVP Decisions: Using the Pillars

During the pilot, you learned the ability to create MVP for a specific result to deliver customer value in the demo. You also used the relative ROI decision map to illustrate the "value vs. cost" and make the best MVP decisions. As you learned, delivery and finance understand the cost, while marketing and channels need to have their fingers on the pulse of customer value.

To improve the MVP decision, the value part of this equation needs to constantly improve. A key to making these decisions is understanding what customers really value. The fuel for a strong MVP decision-making engine is a clear understanding of the customer value. This fuel is dependent on marketing and channels getting better at knowing this answer.

To build on this ability and reinforce it on a larger scale, marketing and channels need to ensure that the customer voice is being heard to make the correct MVP decisions on an ongoing basis. Therefore, marketing and channels need to constantly be looking to improve ways of listening to the customer voice over time. Are there ways to hear the customer by new social media outlets from marketing or new ways to reach the customer from channels?

They key to developing MVP over the long run is to rely on accurate perspectives from the customers on what they see as value. Continuous rigor, discipline, and insistence on hearing the voice of the customer directly will fuel MVP decisions and glean better business results. The peer leadership coalition needs to avoid the temptation of assuming what the customer wants, thus leading to poor decisions on MVP delivery.

BUSINESS CORNER: CUSTOMER VOICE OVER CUSTOMER ASSUMPTIONS TO DETERMINE MVP

I was a senior director of delivery in a market leading company that implemented scaled agile ways of working in the delivery department with tremendous success. One of the things we did was go all in with agile delivery. We brought in the coaches, went through certifications, changed the office layout, and had all the agile ceremonies in place. We even got to a point where we were doing scaled agile ways of working in delivery that were beyond any of our current large competitors. We got the delivery agile thing right and matured to having a fully agile delivery organization. So, we were feeling good about ourselves. We were super-agile in delivery, which meant we were delivering software fast. Our company was very confident in making product decisions since we owned about 90% of the market.

But, like all other market leading companies, we were disrupted by smaller competitors that were agile in delivery *and* the business—agile business disruptors. When we lost our first major client the alarm bells went off, but it was already too late.

What we didn't realize was that speed was only half the answer to competing against disruptors. We were fast, but building products that the customers were no longer interested in buying. How could this be? We thought we knew what the customers wanted since we owned 90% of the marketplace. We were smart! But the lack of real customer intelligence was the fuel that we just didn't have. We didn't have the actual voice of the customer, since our channels, who listen to the customer, didn't participate in our product planning. Although our listening channels had this customer intelligence from their daily interactions, it was not able to contribute or share this with the product planning team. As a result, we had to make educated decisions based on our own best guesses and assumptions of the customer. We made large bets based on these assumptions. Many of them turned out to be wrong. This really caused a huge value chasm between what we delivered and what our customers saw as valuable.

We were forced to acknowledge that we had to change the way we worked in order to compete against these digital disruptors. We went from a steady-state marketplace to one that was changing constantly and were not prepared to sense or respond to these changes.

Meanwhile, those "smaller" agile business competitors began growing because they could sense and respond to what customers would want to buy. They started capturing market share and eventually dominated the space. Not sure if you have heard of them? They were Google, DoubleClick, Yahoo, and Razorfish. My company was sold to another company in 2011.

Continue Working for Small and Specific Results

During the pilot, you will have learned the value of breaking large batches into smaller batches, which allowed the four pillars (delivery, finance, marketing, and channels) to work together to test and confirm product assumptions. This allowed the four pillars to learn what worked and didn't work. Now you can extrapolate those learnings to the business.

How would this apply? The business may have a goal of capturing 25% of a new market within three years. In their business case, they may put in information around market needs, value propositions, product deliverables, and ROI cases. But remember, these are just assumptions based on what they know today. In many cases, these assumptions are just that—assumptions. They need to be tested out.

Waiting until the end of three years to see if we have captured 25% of the new market is way too large a batch. This is risky in today's competitive environment; the market will not stay consistent for the next three years. In order for the business to be successful in this fast-moving market, they need to break down the business goals into smaller experiments with measurable results and be ready to adapt as they learn.

This sets the stage for the executives from delivery, finance, marketing, and channels to become mutual investors in the business initiative: delivery, marketing, and channels are investing time and staff to the proposed initiative and finance is investing funds to support this initiative. They have a common understanding about the business assumptions, as well as what the possible return on their respective investments might be. They provide tremendous value in that it allows the business a way to justify the investment.

Therefore, each customer-facing pillar of the organization needs to adopt a practice of taking those small batches and conducting experiments to test out assumptions. Based on the learnings, the organization can then decide to pivot or continue with their strategies. But instead of these decisions being based on gut instincts, they would be based on the results of good scientific business experiments.

Limiting Organizational Work in Progress (WIP)

Taking a page from Lean and Agile delivery, limiting work in progress or WIP is key to improving the flow of work. This learning is applied here to the broader organizational WIP. Think of your large strategic initiatives in play, you could probably afford to cut them in half and realize a better outcome. Related to smaller batches—shorter projects with incremental results—is the concept of working on fewer things at once. You learned in your pilot that focus helped the teams succeed—focus through the short timeboxes of Scrum or through the work-in-process (WIP) limits of Kanban. But your business leaders don't have that experience.

Instead, they are likely to start everything at once. Delivery, finance, marketing, and channels leaders may hear each other say, "Yes, we have started that project." And that statement is reassuring to them. It's reassuring to know that "I'm at least trying to serve every stakeholder" and that "my colleagues are serving me by starting my projects." As a result, most companies have far too many work items in progress, which is why throughput is low and lead times are long.

But too many initiatives in progress will ultimately will cause gridlock. Why? Studies have shown that large quantities of WIP in fact reduce the amount of work getting done. It's a bit counterintuitive. Having lots of projects in flight *feels* productive. If any project gets stalled, the workers can make progress on another. They remain as busy as possible.

But it's a false impression. Sure, everyone's working. But is the work itself getting the value we're trying to provide our customers over the finish line? The evidence suggests not.

Instead, valuable time is lost in context-switching; every time a person switches gears, they start over, re-entering the context of the second project. Over time, and over many context switches, much time and focus is lost. Not only do the workers waste time, they also experience degraded performance—the solutions aren't as good, the quality of work not so high.

When there is too much WIP, no slack exists in such a system. As a result, there would be no space to think, to innovate, or to absorb change. Any work that takes just a little bit longer has a cascading effect on all the work. Everything gets slower and slower, and the ability to pivot is completely diminished.

The whole system is made worse by the fire drills to show some semblance of progress when some senior leader remembers to ask for an update. The resulting effort to create some "status" (rather than actual progress) wastes time, builds resentment, demotivates employees, and prevents the very innovation and value we seek.

The effect of too much WIP in the system has been quantified again and again. The more work in process, the longer it takes from idea to delivery (lead time). (See Figure 4-3.)

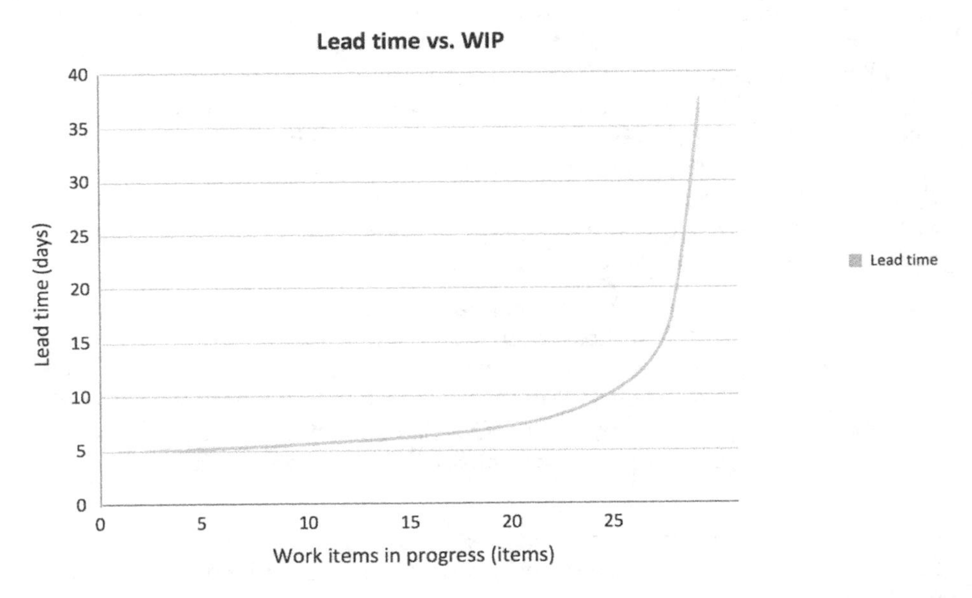

Figure 4-3. Lead time vs. work items in progress

Notice that the negative effect happens suddenly and dramatically. Any given system (organization) has an optimal number of things it can handle. Beyond that point, it suddenly becomes impossible to get anything in a timely fashion. Most organizations we encounter passed this point long ago.

There is a similar negative effect on throughput, or how much work gets done (Figure 4-4). Very little gets out the door. The good news is that we're not trying to find some perfect number of initiatives to put in progress. There's a pretty good sweet spot. Once you stop measuring resource utilization (i.e., whether everyone is busy) and start measuring throughput (the amount of work that gets completed) and cycle time (how long it takes to get a single thing done), it becomes quite easy to see the right number of initiatives to have in flight.

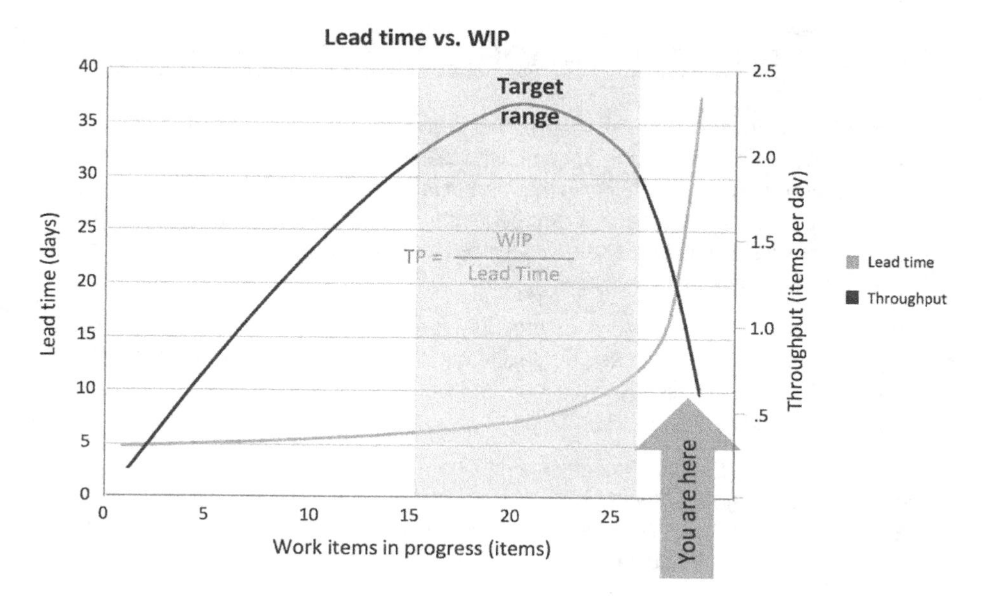

Figure 4-4. Ideal WIP range

Finding the right WIP limit is the easy part. The hard part is the cultural change. In most large organizations, people want to hear their project has been started. They want constant reassurance that progress is being made, even if actual delivery is taking a while. In the context of a traditionally slow delivery engine, no one wants to hear their work hasn't even begun!

Additionally, fewer initiatives means a shift to joint initiatives rather than departmental initiatives, which naturally requires collaboration across departmental boundaries. The real cultural shift is the accountability for outcomes beyond a single department's span of control. Limiting WIP for an Agile team is relatively easy, they simply work on less things, but limiting WIP at the organizational level is hard because it requires collaboration across different departments, reporting lines, and executive reward systems.

It takes real courage to say, "with focus, we'll improve both productivity (measured by throughput) *and* time to value (measured by cycle time). let us decide not to start initiative B until initiative A is done because that will ensure that both A and B get delivered in a timely manner."

And we're not just talking about delivery initiatives. We're talking about all everything to improve business processes from corporate initiatives to departmental initiatives. The organization must do less in order to do more. It's a tough sell, but the payoff is huge.

LIMITING ORGANIZATIONAL WIP SUCCESS STORY

A midsized software company was successfully practicing agile for many years. Around 2014, leaders recognized something wasn't right because the organization was working on a large number of initiatives, but none were delivered. Customers were expressing frustration at the lack of product improvement and feature delivery. The general consensus was that *something* needed to change to improve responsiveness while making a dent in the heavy workload.

The company hired a new VP of product—someone with deep agile and lean expertise and strong leadership. He guided his colleagues to reduce the number of active initiatives significantly, limiting work in process. Everyone was still working hard (harder, really), but on fewer things.

The turnaround was quick. Throughput improved in the very first quarter, where a series of game changing innovations were released to the market. By the second quarter, the delivery problem was resolved, and major product initiatives and innovations were now being delivered to customers on a regular basis. This organization's leadership has continued to embrace limited organizational WIP today.

Given all the evidence, it is clear that limiting work in process at every level in the company has the impact of helping the organization get more done and to be pivot-ready. But it's still a hard sell and a big cultural shift. It must come from the top. If senior leaders are constantly asking "Have you started that yet?," then everyone will scramble to say "yes." Instead, you want every leader to ask "What are you focused on?" and "Are you sure you should start that before finishing this?" To speed up the whole organization, the leaders in each of the four pillars must learn to ask, "Why are you working on that when the other thing isn't done?" and "What will you stop in order to make room for this new priority?" and "How can we descope that project so we can finish it and deliver value to the customer sooner?"

Returning to the related concept of smaller batches, one way that every part of the organization can reduce WIP is to ensure that every initiative focuses in on the smallest delivery increment that will provide value. We work on as few projects at a time as we can, and we make each project as small as it can be while still delivering value. Then we decide whether to build another increment to increase value, or move on to the next initiative.

In this calculus, the return ROI assessment becomes critical. We want to work on only the very few most valuable things. The ROI calculation becomes as much about what *not* to do as what to do.

EASY LESSON ON WIP FOR COFFEE LOVERS

On average, the lead time (how long it takes to deliver) is proportional to the size of the queue (how much work there is) divided by the processing rate (how long work takes).[2] Translating that to your daily visit to the coffee shop, it means that if 20 people (size of the queue) are ahead of you in line at Starbucks and the barista is serving five people a minute (processing rate), you will be served in four minutes (lead time). Of course, this can feel like a long time if you haven't had your first cup of coffee!

To shorten lead time, the shop has two options: improve the processing rate or reduce the work in process. Usually, we get to an optimal processing rate pretty quickly; after all, it does take a while to get a caramel macchiato just right. So the only way to shorten the wait time is reduce the number of drinks being made at once by the same barista. The other thing that limiting WIP does is favor completion over activity. Imagine the barista filling a hundred cups one drop at a time. There's a lot of activity yet a hundred customers are unhappy, but in the end it's better to deliver small batches of actually hot coffee.

Create Flow at the Organizational Level with Smaller Batches

In order to achieve true velocity, all four pillars of an organization need to collectively identify the most impactful initiatives and also collectively limit the amount of work while breaking down the work into smaller batches. But in today's fast-paced business environment, that may seem counterintuitive. In many organizations, each of the four pillars may be tempted to take on as much as possible, and as big as possible, thereby putting collective initiatives at risk. The key message here is to shift our thinking from the big blue-chip initiatives, to faster "test and learn" MVPs or minimally viable product delivery. This is how we have smaller batches, which enables faster flow of value through your organization and into the market.

If all four pillars of the business collectively focus to complete the most impactful initiatives, the following benefits can be realized:

- Decrease in time between idea generation and business outcomes. Business outcomes cannot be realized until the customer has the product in their hands. This can only be accomplished if all four pillars deliver value in a coordinated manner. The business needs to identify the opportunity, delivery needs to deliver product, marketing needs to promote it, and channels needs a way to get it to the customer.

[2] "Little's Law" is a fundamental of queue theory and defines the relationship between WIP and cycle time. It is the reason why teams try to limit WIP.

- Reduction in wait time between groups. For instance, if each pillar can avoid waiting for a large batch of completed work, small units of work can be passed through each pillar to accelerate overall delivery.

To help illustrate this point, in the book *Lean Thinking* (Free Press, 2014), James Womack and Daniel Jones recount a story of stuffing newsletters into envelopes with the assistance of the two young children of one of the authors. The task was quite simple but had many moving parts. The multistep process involved addressing, stamping, stuffing a folded letter, and sealing the envelopes.

They divided the pile in half, and ran a challenge between the children versus their father to see which team finished first. Seeking efficiency, the children opted to fold all the letters, then attach all the seals, then place all the stamps. Meanwhile the father chose to complete each envelope at a time, with all the involved steps. The father won the race. He was able to complete the full cycle of work from start to finish. Tying this back to our business context, there are always time milestones: a market entry date, a tradeshow date, regulatory compliance date, etc. It is wiser and, as it turns out, more economically viable to get the complete valuable products and features out to market before the milestone hits. Or as the story goes, more completed envelopes, rather than building a massive inventory of unfinished envelopes.

The ultimate goal becomes one of flow. Rather than optimizing locally for one of the pillars, organizations can collectively focus by breaking the value into smaller pieces that can go out to market faster, velocity will naturally increase.

"Cadence Pivot" Meetings

While the quarterly planning meetings will formulate the overall business strategy, the cadence pivot meetings will take this strategy and hold the executives accountable on their commitments to implement the strategy.

Cadence pivot meetings are based on the customer-facing pillars being held accountable to each other: delivery, finance, marketing, and channels. Cadence pivot meetings track each customer-facing pillar's actions and progress toward the goal of delivering customer value in a highly transparent way. It is imperative for delivery, finance, marketing, and channels to collaborate and make the right decisions on bringing new products to market.

Cadence pivot meetings are where you introduce the concept of "pivoting" to really deliver customer value (Figure 4-5). Today, high demands from customers require that companies replace the old rigid processes with a readiness to pivot in response to demands. Pivoting gives life to an agile business and makes it possible to sync the customer-facing pillars in a way that has customer value at its core. When companies can pivot, market changes and new customer demands are no longer feared, but rather welcomed as an opportunity to outmaneuver a competitor.

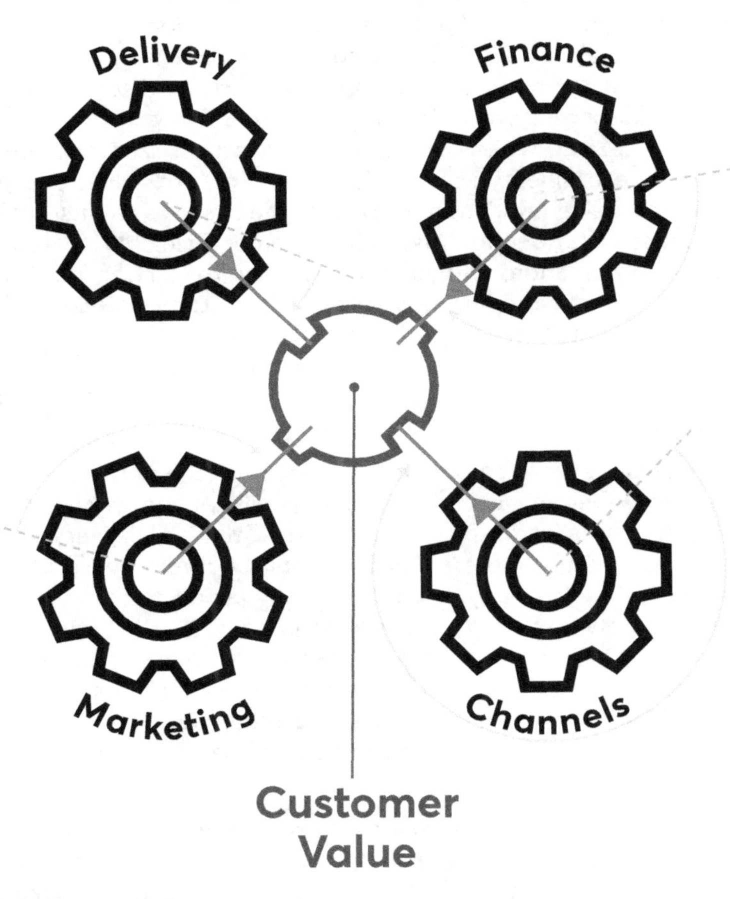

Figure 4-5. Pivot toward customer value

Executive Accountability to Deliver Business Results

During these cadence pivot meetings, all eyes are on delivering customer value and achieving business results. As a result, executive efforts and decisions are made to continue on the current plan, stop activities, or pivot to a new direction to achieve this agile business realization. The pillars meet and actually make decisions, many times difficult, on the fate of the software in play.

The company executives meet as a peer leadership coalition of delivery, finance, marketing and channels for half-day pivot meetings, typically biweekly or monthly. These meetings will do a retrospective of progress and provide feedback to plan for appropriate improvements.

The cadence meetings are vital to achieve agile realization. Since agile business realization is not about delivering product, but rather delivering business results, these cadence meetings are the mechanism to keep track with full visibility and transparency to hold the peer leadership coalition accountable for these results. Since we know that agile realization can occur only when you measure the business results of your actions, not when you measure simply what was delivered, it's about revealing the business results, not delivery results. Cadence calls now hold the four customer-facing business executives to have these conversations with full transparency. To start, finance, marketing, and channels will share the business results and customer feedback from the current product deliveries. They are all held accountable for the business results from their respective areas. The team is held accountable by their own executive peers, as a community, to make changes based on those experiments. It's now about accountability to deliver business results.

Without this, delivery is left alone to try to measure the only thing they know, which is delivery. Delivery-based metrics have always been focused on "How we can be better at delivering products?," because that's all delivery knows how to measure. As a result of these delivery results, delivery delivered products fast, speeding up the mess, by delivering wrong products that the customer didn't value. It's now up to finance, marketing, and channels to fill that gap and change the questions to "How do we make a customer's experience better and glean better business results?" Those are the new business results metrics we need to focus on during these cadence meetings. The customer voice from finance, marketing, and channels must feed that delivery engine to build the right product.

The full-transparancy gained enables "cause-and-effect" evaluations based on business results. They are constantly asking questions and tweaking their decisions to answer the ongoing, infinitely repeated question of every cadence meeting: "What happens to the business when we try this, and what happened to the business when we tried a similar experiment in the past?"

With this new transparent, results-oriented cadence meeting culture, the conversations hold these peers accountable for decisions. When one of the customer-facing pillars makes a claim that they want delivery to build a certain product, delivery can now hold that pillar accountable for the business results. So, if channels says this product must be built, delivery can ask how much revenue they you willing to commit: "If we build this then can you commit to that?" The online transparency internal negotiations are the key to peer accountability from within the customer-facing pillars.

Delivery Accountability to the Cadence Pivot Meeting

Since delivery will be in the hot seat to develop more compelling customer-valued software, they will be asked to host the regular cadence of pivot meetings with the other three pillars: finance, marketing, and channels. By taking a leadership role in these pivot meetings with the other pillars, delivery will have an understanding with current information to build software that the customer actually values.

They will measure success based on how much customers use the functionality that is built into the software. The higher the usage, the more successful delivery has been.

Finance Accountability to the Cadence Pivot Meeting

Finance accountability to the cadence pivot meeting will accomplish the following:

- Provide short-term fund planning and avoid detailed long-horizon planning. This is a very different approach, since finance traditionally expects a long horizon, budgeting annually with detailed cost estimates that must be frequently updated, often quarterly. The new approach is designed to embrace change and uncertainty: the final product is being impacted based on what everyone learns during these sessions. Finance must learn to embrace variance over accuracy.
- Set out to do work based on the priorities set by the pivot meetings. This is different since finance project cost accounting requires a reapproval for any delays that increase the project budget. Rigid approval limits pose an unnecessary control and delays caused by the budgetary approval process.

Finance will be asked to participate in the pivot meetings with delivery, marketing, and channels to create a new pivot-enabled funding model, which will directly impact the business pricing decisions in a positive way. Better, higher, and more profitable pricing starts with a change in the way that finance funds programs. This means that finance will be asked to fund products incrementally (quarterly) within a regular cadence, instead of the traditional big-bang annual funding decision-making approach.

To support this new iterative funding model, finance will be asked to create a deferred decision-making model that is supported with a resource pool of funding. Deferred decision-making means that finance can rebalance the investment portfolio based on the new changing priorities that that are learned from the other three pillars.

From this lump sum fund pool, finance will spend in small, incremental amounts and pivot based on customer value determined during the pivot meetings and as new insights are uncovered.

Finance will, therefore, transform their focus away from their department goals and adopt a new focus on customer value. With this new focus, instead of the big up-front investment planning, finance will be funding based on customer value and potential profit margin results. This means a shift away from annual planning and funding models to these incremental cadence times. These pivot meetings have finance working alongside the other pillars to make incremental funding decisions. Finance will meet frequently and have the authority to make quick decisions to invest in the most valued and most profitable software going forward. Furthermore, finance will focus on the highest-priority and most customer-valued funding objectives.

They will measure success based on how much profitability the software generates. The higher the profits, the more successful finance has been.

Marketing Accountability to the Cadence Pivot Meeting

Marketing will be asked to attend a cadence of meetings with delivery, finance, and channels to alter promotional and branding messages based on customer value feedback from the market. These cadence meetings will focus marketing on responding to market changes rather than on following a plan. To do this, marketing will be making smaller more frequent decisions based on what is learned during each session instead of one big-bang, best-guess prediction.

They will measure success based on how many leads convert to making purchases using the software. The higher the conversion rate, the more revenue, and the more successful marketing has been.

Channels' Accountability to the Cadence Pivot Meeting

Channels will be asked to share customer intelligence and provide input on what customers want during the cadence pivot meetings in a continuous loop with delivery, finance, and marketing. Channels will need to commit to sharing customer feedback, learnings, and stories based on customer value. This alignment will help make sure that channels understands the root of why these offerings are being funded and delivered and what customer need is being addressed. This should help our channels to be more successful because they should have greater empathy with the customer's challenges and how the offerings address it.

Channel members provide additional feedback around the resonance of the messaging and whether the new product/offering is actually addressing and delivering the intended value. This provides another conduit in helping ensure you're delivering the right offer to the right customer at the right time and for the right price.

They will measure success based on how much they can increase sales from the software improvement ideas they provided. The higher the sales, the more revenue, and the more successful channels has been.

Meeting Agenda

These cadence pivot meetings create tactical experiments to test which ideas are important and which ones are the riskiest. The result will enable better decisions about the MVP from the leadership coalition.

These meetings have some basis in "lean canvas"[3] concepts, which are matured by updating the canvas to hold the customer-facing pillars accountable to achieve business results. During these cadence pivot meetings, the leadership coalition will implement the following standard agenda.

Around the Horn

Leadership coalition executives from delivery, finance, marketing, and channels will take turns providing answers to two questions: "What did we learn in the last period?" and "How, if at all, did my own pillar contribute to realization of the desired business results?" These questions not only run across each pillar but also cut horizontally across the business; in addition, the respective pillars will be held accountable for answering the following questions:

- Delivery: What does the product look like (show demos)?
- Finance: Have we met profit projections?
- Marketing: Is the lead quality improving?
- Channels: What are customers saying about this?

Transparent Collaboration

During these cadence pivot meetings, it's important to create a collaborative environment that fosters the necessary safety for productive debates and one that encourages each of the pillars to challenge each other. These meetings depend on the courage of the executives to speak up and share their views, even if others don't agree. Otherwise, this becomes just another status meeting where people drone on and everyone agrees and leaves.

[3]**Lean canvas** is an adaptation of **business model canvas** by Alexander Osterwalder, which Ash Maurya created in the **lean startup** spirit (fast, concise, and effective startup). **Lean canvas** promises an actionable and entrepreneur-focused business plan. It focuses on problems, solutions, key metrics, and competitive advantages.

A word to the wise: these collaborative discussions can be emotional because many executives might want to protect their project for personal job security. The leadership coalition, therefore, needs to foster and structure an environment that encourages open communication to make accurate observations and the correct decisions based on delivering customer value and business results, not personal job protection.

Decision Discussion to Answer Questions

- Keep - What is working? Continue or increase program
- Kill - What is not working? Divest or hold program
- Pivot - What can improve? Create a new approach

These are vital decisions to keep, kill, or pivot projects.

It's important to reinforce that killing or pivoting on a project is not failure. Small batches and frequent review from these cadence meetings help to build that message, because the sunk cost of a few weeks of development is relatively low. A lesson that a product is not performing is just as valuable as learning from successes. It is better to see each small batch as an experiment, even if it does not deliver the desired results, it should be assumed it was an experience to learn about the market and avoid bigger expenditures. Then, that experiment should be mined for useful information. Underperforming products provide valuable information to make decisions to kill or pivot. However, the myth of the "sunk cost" is strong in corporate culture decision-making. Everyone can agree that they should stop developing that failing product, but they still hold onto them because they believe they have already invested so much. This sunk cost myth can be a barrier that keeps the organization from finding the next big bet by pivoting or killing the dead-end products. The good news is that the cadence meetings take place precisely to avoid these massive sunk costs, since decisions are made every few weeks, not over months or quarters, for which costs are larger. So failing fast, in a few weeks, is cheaper than stubbornly dragging out development on a product that is not delivering business results over many months.

COACH'S CALL-OUT: THE IMPORTANCE OF FAILING AND FIXING FAST

Rob Desmarais, Senior Director, Agility Services

In a corporate setting, many people think failure is a bad thing and will hurt their career; when working with an agile mindset that could not be more false, albeit with one caveat: fixing or improving what can be done better. I personally see it as an excellent quality when someone can identify where something is broken and have the courage to stop and fix it as opposed to going through the motions until it's too late or until the hero mentality saves the day. This story shows the importance of identifying opportunities fast and how taking the time to fix them early can significantly help your organization.

I was working with a client on a large, multiyear initiative to streamline the customer experience and consolidate/modernize a high number (>500) of legacy systems consisting of approximately 500-600 people from different parts of the organization. One of the first sessions was an offsite event with initiative leadership to identify the team structure at all levels (portfolio, program, and teams) and foster cross-functional teams as opposed to component teams. As an outcome of that session, two options were identified, and ultimately the customer chose the option not recommended by the consultants.

As a next step, the program management team identified the MVP, did preplanning refining and prep, and had a successful first planning session. Two sprints into the first delivery increment, the teams started complaining that there were too many cross-team dependencies that were blocking progress and significantly impacting the efficiencies of all the teams. This lead to a session with the leadership team on the importance of inspecting and adapting early and often so problems don't continue to boil over. In agile it is very important to identify failures early and fix them immediately, as opposed to waiting until it's too late and deadlines are missed or the budget is blown out of the water.

In a legacy approach, this process could take weeks or even months while the teams continued to work on potentially the wrong deliverables and continue to burn through the budget. Instead, the outcome of our session was to put everything on hold, realign the teams in a more efficient way where teams can fully deliver units of work by minimizing cross-team dependencies, clarify the MVP in terms of reducing scope and therefore WIP, and redo planning. You may be wondering what the team members did during this "hold pattern": their time and cost were not wasted. This also allowed the team to focus on identifying architectural and infrastructure-related work that could be done during this time instead of later in the process. The time was also used to help set up additional automated testing infrastructure to streamline the testing in the future.

While some executives questioned this, they trusted the leadership team and approximately three weeks later conducted the planning again. The outcome of this hold/restart led to a successful pivot that established a team structure and framework for the ultimate successful delivery of the MVP ahead of schedule and under budget.

This leadership coalition will need to support the difficult decisions so the executives feel like they have job security if their project are canceled. This means there needs to be an organizational structure that allows the ease to move on to other high-value projects without a risk of losing their job. There needs to be a new organizational alignment with this new way of working that makes it safe for executives to kill projects of low value and then reallocate their time and focus on a new high-value project. The organization might be aligned differently to move high-value work to the executives wish ease. This is creating a new organizational alignment that fosters open communication during cadence pivot ceremonies (more on this in the next section, "Changing the Organization Culture") (Figure 4-6).

The Cadence Pivot Meeting

Keep? Divest? Pivot?

Each representative from the four pillars is responsible to the group when making decisions about keeping on course, pivoting, or divesting. The first set of questions that have to be answered are: "What did I learn in the last period?" and "How, if at all, did my own pillar contribute to realization of the desired business results?" Other topics to discuss are:

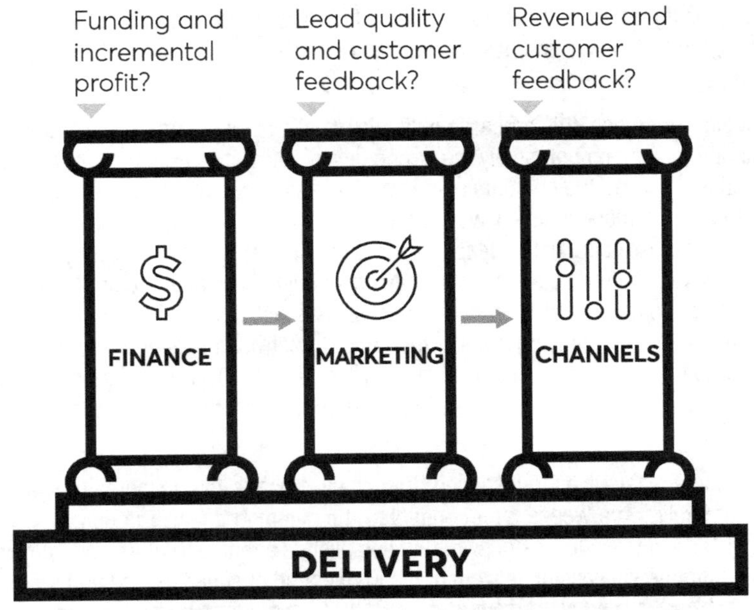

Figure 4-6. Cadence pivot

COACH'S CALL-OUT: TO PIVOT OR NOT TO PIVOT

Chris Browne, Senior Director, Agility Services

The ability to sense and respond to changing market conditions or competitive threats is essential in today's fast-paced environment. By becoming agile at scale, you can enable your organization to respond quickly to fast-growing competitors and market disruptions.

This in turn raises the question: "Just because you can pivot, should you?"

The danger of not adjusting course is that a more nimble competitor can capitalize on a market opportunity quickly and by the time you turn your ship you're already too far behind to catch up.

By pivoting too often, you run into an equally significant danger. When we change direction just because it's easy to do, we run the risk of never actually getting anything to "done." Although your teams may be busy, the reality is you're thrashing without delivering anything of value.

When we initially established our quarterly steering and big room planning cadence at Rally, we struggled to deliver anything of significant value to our customers. Since we now had our planning, development, and delivery rhythm synchronized, we were able to change course frequently. Unfortunately, this led to "shiny object" syndrome, where we frequently began work on features that were new and exciting or easy to get out the door. Although we had a lot of small wins, we also had many big misses. As new ideas came in, we stopped working on our current set of features, which resulted in a never-ending backlog of features to be finished.

Feedback from our customers was great initially. Their special requests and top features were making it into the product quickly. Over time, however, their feedback became more and more negative as we neglected to meet our commitments to complete many of the half-baked features in our product. During this time, our competitors were able release functionality that brought them much closer to feature parity and became more competitive in the sales cycle. The best part of agility is that you're always learning; luckily we received feedback quickly, which allowed us to change our processes and mindset.

The decision to pivot is a constant balancing act: you should check in regularly, validate whether or not you're on the right course, and only steer if necessary. When your regular steering meetings feel boring because you aren't making too many changes, that may very well be a good thing.

Changing the Organization Culture

In the introduction of this book, I described how agile thought leaders are often most challenged by the idea of changing a culture. Well, now, with all the momentum from the previous three strategies executed, and the agile business realization plan in place to support the new operating model, you can finally strive to accomplish organizational culture change. You've tackled the process with the agile business realization plan, but now you have to take on organizational change.

However, humans have feelings and emotions that can become forces against agile business adoption and that may be a roadblock. Specifically, people may feel:

1. Marginalized, redundant, and threatened when projects are canceled. If team members consider themselves special or important, they will experience a personal loss. When making decisions to kill programs, it will be difficult to be successful unless the delivery can move high-value work to areas that have capacity. As a result, the organizational structure can bog down the product decision-making process in a tangled web of endless debates, since executives cling on to lower-value projects to support their job security. As a result, a self-preservation protectionist mentality emerges that gets executives stuck on defending their own projects to keep them going at all costs.
2. Exposed due to the new transparent environment and needing to understand each other as a leadership coalition. Since, naturally, each side does not fully understand how the other operates, they can talk past each other on issues. For example, finance, marketing, and channels will not know, and should not be expected to know, delivery resources, or how many software developers it will take to create a specific software function. In turn, delivery will not know the detailed tactics of an overall business strategy or vision. Therefore, they need to find common ground. As a result, a finger-pointing mentality emerges that undermines trust and hinders communication.

Since we are dealing with people, not process, now is the time to get HR involved. The HR department of the company can play an important role in addressing these people-based concerns. Engage HR with the intent to make changes to the organization to support the new agile business approach. It can be a strategic partner in establishing an agile business organization structure.

To support your reinforcement plan and change the organizational culture, HR can help to

1. Establish new metrics to drive change
2. Create a product management office

Establish New Metrics to Drive Change

Create measurement mechanisms that are aligned with the same mutual goal of delivering customer value. Even with all the best intentions from executive peers aiming to follow and sustain the changes put in place, there may be a natural gravitational pull away from change because of how the organization has previously been structured. The four pillars ultimately have been traditionally measured in terms of department goals, but this method of measurement is disincentivized by the customer value focus of this executive peer group. So the challenge is how to get them started and how to shift their mindset from apprehensive to motivated.

The answer is based on how these executives are measured for success. In short, measurement drives behavior in the long run. In most organizations, delivery, finance, marketing, and channels are measured as individual departments and, consequently, the success of one pillar is irrelevant to the other. As a result, instead of working in unison, quite often they work disconnected and sometimes even in conflict with one another. While that could be seen as business as usual, unity across pillars can actually provide you with a competitive advantage. As William Edwards Deming states, "A company could put a top man at every position and be swallowed by a competitor with people only half as good, but who are working together."

So how do you get started? The first thing is establishing unifying results that can be measured across all the pillars. As Deming suggests, "most often the efforts to have people collaborate is difficult when the primary focus has been on optimizing what they are individually and departmentally accountable. The organization as a whole can suffer even while improving the results of separate department components because the most significant gains are to be made in managing the company as an organizational system." Common results will serve as guiding stars for all areas of the business to help inform how they maneuver and deliver customer value in the shortest time possible.

Having results is a great start. But you also need the ability to measure progress. As you design your measurements, you need to consider the positive as well as the negative effects of what you measure. There is a well-known concept called the Hawthorne Effect. The Hawthorne Effect is a psychological phenomenon that produces an improvement in human behavior or performance because of increased attention from others. How executives are measured for success will drive behavior. Therefore, when we create measurements, we need to make sure they drive the right behavior. So even with well-intentioned results, bad measurements can lead to failure.

Let's take a look at an example:

- Business result desired: Increase market share by 25%.
- Goal: Increase customer satisfaction by delivering at a faster cadence and with high quality.
- Measurements: How much code is written and how many bugs are fixed?
- Result: Developers want to hit and exceed those measurements, so they end up rushing and checking in lots of buggy code, which causes many defects, which they get to fix.

In this case, the measurement of more code and more defect repair was probably met, but the unintentional result was that the customer received a poor-quality product.

So, what would be an example of a good measurement?

Let's look at the Oakland A's baseball team in 2006. They ranked 24th of 30 major league teams in player salaries but had the fifth-best regular-season record. How did this happen? They had a team that played had a common, specific, measurable goal, which was "on-base percentage," which refers to how many times batters got on base. They chose this one metric to serve as the team's leading indicator of success even though there were hundreds of stats the coaches had at their disposal.

The same reality applies to organizations today. Like the Oakland A's, the customer-facing pillars need to act as a team and cut through the noise of the multitude of metrics options to hone in on a key result that they all agree will help them succeed. That commonly measured result is customer value delivered. We need the customer-facing pillars to be able to all work in unison, agreeing on how to achieve this common result goal.

The first principle of the reinforcement model is for the company to take an organizational view of customer value as opposed to a departmental view and to bring the same principles that brought our development and quality assurance organizations together toward a common aim.

HR will be key. HR needs to develop a model for developing performance metrics that sustain the agile culture in the organization. Although there is no massive reorganization required, HR will need to support your efforts and align how executives are measured and compensated for success. Including core responsibilities, HR needs to include "new responsibilities" or KPIs.

Create a Product Management Office

In your company's journey toward organizational change, you have just nailed down the process changes, which include the customer-centered operating model that you have proved during the pilot and your reinforcement plan. Now comes the "people" part, which is about culture change. If you want to implement change successfully, you will need to have the right people in place to build the change ecosystem around the new operating model.

Getting the right people and culture is tough, but you need to have a change agent—a set of players that mobilizes others toward change. To do this, the leadership coalition will establish a "product management office" (it can also be called "product strategy office"). While this is a new office, it's an elevation of product strategists to a higher and more prominent role. They will have extended responsibilities as the liaison to the customer and across the pillars. They will drive the four pillars to make the correct product decisions based on the highest customer value and will find the right balance of business vision while respecting the technical realities of delivery and architecture resources and other constraints. This decision-making office walks a fine line and has full credibility since it understands both business and delivery, with the common goal to deliver customer value.

As you can see from Figure 4-7, when this is successfully implemented, the picture of pivoting evolves from what was described earlier in this book. In this new pivot picture, the customer-facing pillars converge and the ellipses overlap in the middle over customer value. The product management office drives the customer-centered operating model collaboration to the point where there is a blending of organizational lines under the common goal of delivering customer value. The hard organizational lines become softened and blurred with this common customer focus.

Figure 4-7. Product management office

This new office is unique in that it spans both delivery and the business. As a result, this office will understand and have empathy for delivery and its technical capacity and realities while balancing this with the prioritized customer value–oriented product decisions that deliver the highest customer value. Given the unique empathic skill set of balancing business goals and technical constraints, the natural evolution would be for today's product owner to head this office and be promoted as the chief product strategist.

There is a lot of variation in how to formulate this office inside companies. It can come from one of the dominant pillars based on the company. For instance, some companies that are very marketing focused can have the product management office coming from that department. In any case, it should incorporate the *project* management office (PMO), since the executives in this department typically understand both the delivery technical side and the business side, which is key for cross-pillar communication and translation. For this to work, the PMO will have a new mandate. Very often, PMOs are staffed by people who are trained to look for ways to control the project. Therefore, if a PMO is going to evolve to be able to look at business performance instead of project control, then they need to agree to this. This will require reorientation to the goal of delivering customer value to achieve business results as a prerequisite. If the candidate cannot make this new commitment, the leadership coalition will need to select a different candidate that meets that criteria.

The new product management office is responsible for enabling the company to sense and respond. To accomplish this they must do the following:

Enable Sense-and-Respond Decisions

The product management office leads the decision-making process from working with the four pillars. They are responsible for two things: first, holding marketing and channels accountable for providing the right intelligence from their observations in the marketplace as the antenna to "sense" customer desires, and second, making the right decisions to "respond" and make the right products. To do this, the product management office fills the following roles:

1. Represent the voice of the customer for the delivery organization.
2. Conduit for synthesizing external market and internal delivery signals to maximize business value, with the ability to understand and process external market and internal delivery signals.
3. Visionary and owner of the product roadmap. Coordinator for a smooth flow and transition toward high-value work as executives move with ease off of low-value work that has been canceled, so they always have something to work on of high value.
4. Translator and facilitator who speaks the languages of delivery, finance, marketing, and channels. Receives the signals from marketing and channel pillars on sensing the market.

The product management office is the liaison between customer voice and business and the product development team. They are the ones that can drive cross-collaboration and bring the multiple voices together to make decisions about the product going forward.

While an organization needs to position itself for change, it also needs to have a thoughtful approach. At an organization's core lies it's True North, the values that it has, as well as the customers that it serves. Think about Apple as a company and what they stand for. They feel that they exist to challenge the norm and everything they bring to market reflects that value. Think about the new iPhones that no longer have a headphone jack; they wanted to challenge the norm of having to have wires to connect and believe that there is a better experience out there. At first, there was a question in the marketplace, but quickly thereafter many began to adopt Apple's perspective as their own.

I use this example to highlight that creating a product strategy is an art as well as a science. When an organization delivers a product into a marketplace, it's not just about delivering new features and functionality. If that were the only thing, an organization could quickly fall into the trap of getting into a feature war, and as we know feature wars are very costly and will not make an organization successful.

Instead, as we think about product strategy, we need to be thoughtful. Is what we are bringing to market in line with our organization's True North? Is that what our customers value? In addition to internal factors, external factors also need to be considered. What are the trends that exist in the marketplace, and will those trends be disruptive to the business? For instance, think about what Box.com did. When Apple announced the iPad, within hours they were working on an app. Did Box know that the iPad would be a success? Probably not, as at the time of the announcement not one iPad had been sold, but they were able to sense and respond. The iPad was something that Steve Jobs fully supported and that he touted as the future of tablets. Based on who Steve Jobs was, as well as Apple's successes, the folks at Box.com made an educated guess that they needed to deliver an iPad-based product. And they were right.

The Product Management Office Structure

The product management office works best from evolving and scaling the product owner role. These product owners already concentrate on defining the functionality of products to deliver.

In the new, elevated product management office, the chief product strategist is someone who decides what to deliver and in what order, relative to the other initiatives in the book of work. The role of the chief product strategist is to gather and prioritize the various sources of information from the pillars during quarterly steering and cadence meetings: marketing—tracking social media comments, and channels—getting feedback from the sales force in the field. The chief product strategist is at the top level, gathering intelligence from the customer voice from the various colors, and makes strategic high-level decisions on long-term initiatives (see Figure 4-8).

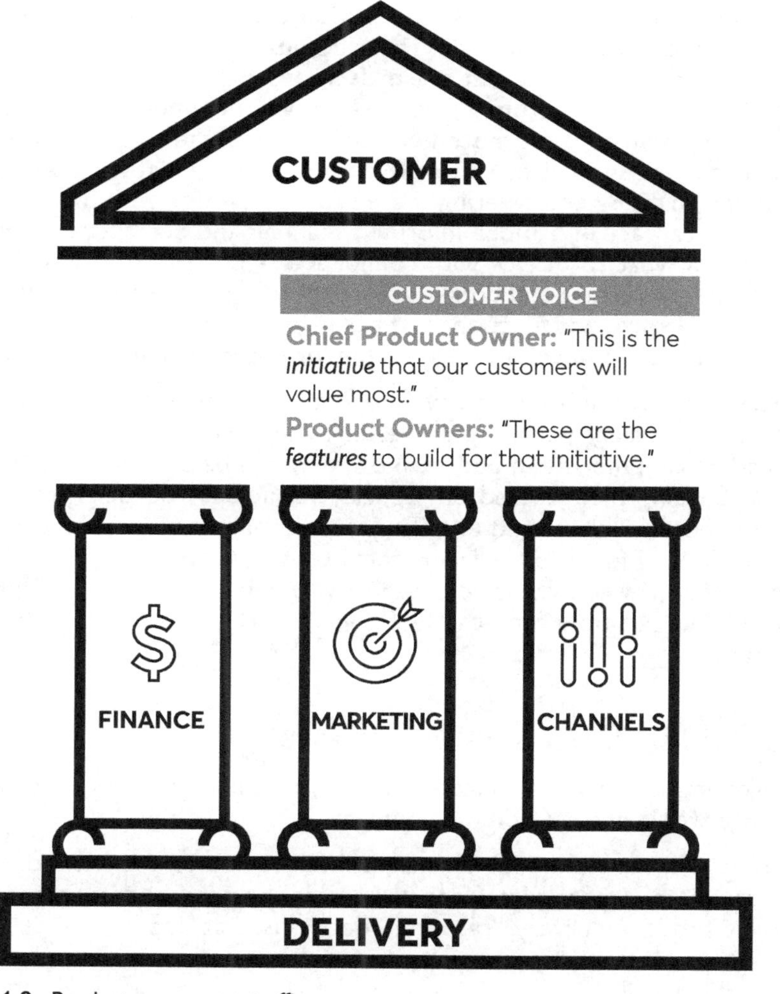

Figure 4-8. Product management office

The product management office will drive consistent refinement methods for each of the quarterly steering and cadence meetings. These will include the following:

- Cross-collaboration of participants needed for the product in all meetings.
- Participants from the business, customer or customer voice, QA, test, developers, architects, systems, and so on: whatever is needed to define cross-collaboration for a specific product.

As such, it is important that the product owner team straddles delivery and the business, and is someone who is action oriented. Product owners will enjoy a new sensation: confidence. That's because they are no longer guessing what to build, but instead can make decisions based on the highly valued intelligence from the customer's voice. They will be empowered a new way because they have accurate customer intelligence on their side. They will have the fuel to create the right software from the best initiatives based on the customer's voice. By empowering these product owners with the customer's voice, they can act in a more informed manner and stand strong on their decisions on what to deliver with confidence. The current product owners concentrate on refining and breaking down this large initiative to determining the smaller features and functionally that can be developed within the cadence meetings. Product owners are responsible for putting initiatives into action.

Delivery is simply the software development teams, including project manager, engineers, and quality assurance and testing. They will be responsible to deliver the product on time within budget. Now, they can be confident that they are not just building things based on a guess, but actually building product that the customers will value and use. They are building the right thing. Knowing this is motivating for these delivery teams, who have for so long been building products that may or may not be used or valued by the customer. They were building blind, but now they can feel confident they are building something important.

Respond (or Pivot) to What Is Sensed

The product management office has the responsibility to sense and respond to change within and outside the company.

A dotted line from the product management office to the head of the the business unit or departmental lead will enable the product managers to make the right product delivery "respond" decisions that impact the delivery of customer products that achieve the desired business results. This dotted-line accountability will connect the leadership coalition within a newly established office called the product management office to the critical decisions made with the pillars. Another purpose of the new dotted-line leadership is to let executives feel safe to make decisions and better collaborate as a peer leadership coalition.

Being Pivot Ready

Pivoting enables a company to quickly sense market opportunities or threats and respond by delivering products that the customer values. Customer value can also mean a positive customer experience, ease of use, or desired product features. To keep up with disruptors in today's application economy, there is only one way to accurately, effectively, and consistently pivot and deliver customer value: the customer-facing pillars—delivery, finance, marketing, and channels—must work in synchronization. To compete against digital disruptors, your organization depends on these customer-facing pillars to clearly sense and understand the changing customer demands and respond and deliver what they want.

As a result, the company can deliver better software to the market quickly, adding value for the customer and bolstering the company's profitability. Today's business is all about speed, and being able to pivot enables this agility.

Agile Business "Pivot Enabled"

When Channels Sense and Pillars Respond (Figure 4-9)

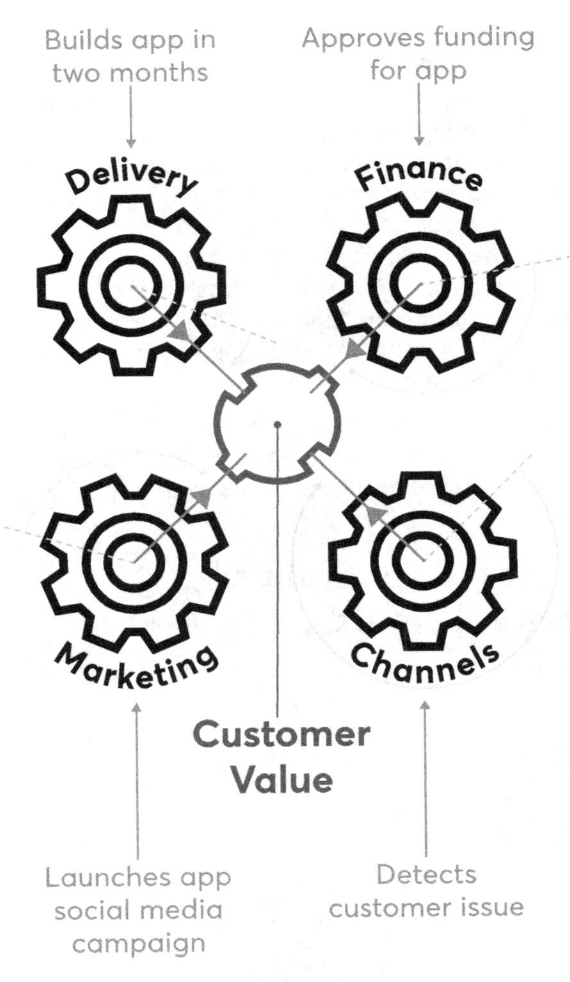

Figure 4-9. Channels sense and pillars respond.

Channels receives information from salespeople in the field, reporting recent sales that were not closed. In addition, consultants report that customers were not happy with the some aspects of a particular product. The channel shares this customer voice intelligence with delivery, finance, and marketing. Together they decide to "pivot" and come up with an innovation, such as a new app, to resolve this issue and give customers what they are asking for, plus something new that the competition does not have. Channels goes back to sales and consultants in the field to validate if this app idea will satisfy their issue. Both sales and channels come back with reports that customers would find this app idea valuable.

Marketing tests this new app concept with social media and receives positive comments. Upon confirmation that delivery can build the app, marketing prepares to launch a new social media campaign.

Delivery confirms it can build the new app idea from channels and deliver it in two months.

Finance understands the revenue ramifications of the app and approves additional funds.

The response of "marketing sense" and the pillars is shown in Figure 4-10.

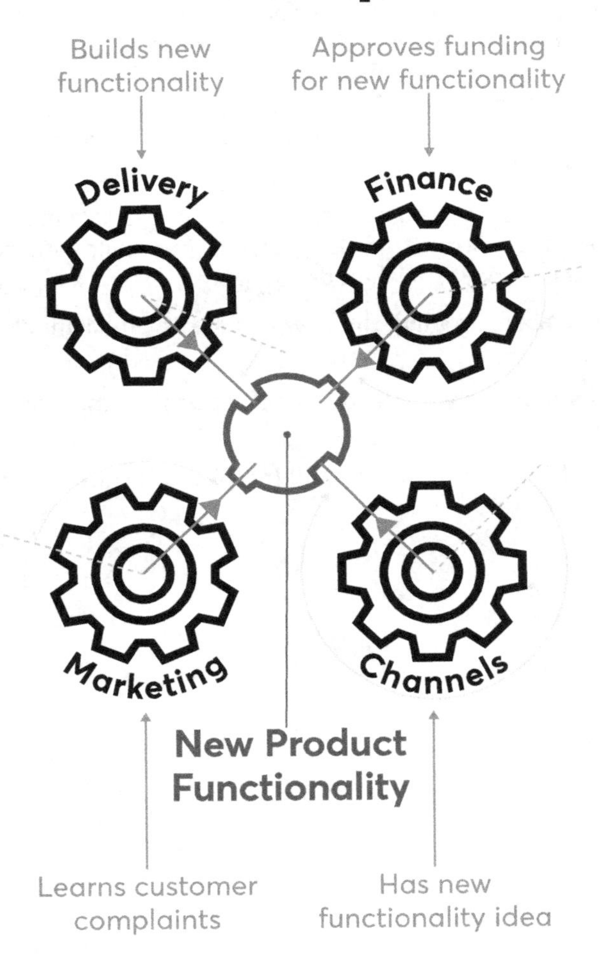

Figure 4-10. Marketing senses and pillars respond

Marketing begins noticing negative comments and customer complaints on social media about a particular product. They track this as a potential problem that is beginning to form a negative reputation and share this with delivery, finance, and channels. Upon collaboration with the other pillars, marketing prepares a social media offensive to answer the negative comments with a solution.

Channels has customer service validate that the reason for the recent complaints is customer frustration from a desired functionality that is missing. Together, they decide to stay on course and persevere since a pivot was not required in this case. That's because delivery was already on track to create and launch a new functionality add-on with additional capabilities that would solve those complaints. Upon delivery confirmation that it is building this on schedule, channels prepares the customer service representatives to share this information with customers to give them confidence that the problem will be resolved.

Delivery confirms that it will continue build the new innovation add-on from channels and deliver it in one month.

Finance has already approved additional funds required to create the innovation add-on to retain customers.

Stay the Course

Remember the story about Netflix as a disruptor of Blockbuster in Chapter 1? Well, Netflix kept the pedal to the metal and didn't let up after they were on top. In fact, they continued to push their company to sustain its ability to pivot after they knocked off Blockbuster in 2013. Without their ability to continue to pivot quickly from a business perspective, they wouldn't be dominating the video-streaming industry today. Although mail-order DVDs were their formula for beating Blockbuster back then, look at where they are today. Who would have thought that a mail-based video rental company in 2013 could be one of the world's experts in streaming technology which other companies are still scrambling to understand? Because they had a customer-centered operating model, they continued to pivot. Therefore, they were able to change their entire business again and are now a giant in Cloud technology. Netflix continues to evolve and your company can succeed as well.

Let's Review

1. Agile business realization is about delivering the results of customer value
2. The reinforcement factors to achieve this are as follows:
    - Establish a new leadership coalition
    - Deliver customer value at a reinforcement cadence that holds executives accountable to their peers
    - Create a new organizational alignment to support this change

PART

2

The Story - Strategies in Action

CHAPTER

5

The Solar Corona Insurance Company

Throughout the chapters in Part 2, we will be looking at the fictional breakthrough stories of several delivery and business leaders who knew very little about agile: we'll combine them into one composite persona for this book, a leader we'll call Linda. Linda is just like you and me—a conscientious and respected executive in her corporation (Solar Corona Insurance Company), trying to succeed with the new assignments given to her. As you will see as the story unfolds, she made strides to achieve business agility, but for every two steps forward, she took one step back.

First, let's set the stage by looking at Linda's company, the Solar Corona Insurance Company. Next, we'll get to know Linda herself. At the end, I'll lead a review session as Linda's agile consultant.

Special thanks to Gene Mrozinski, Advisor, Transformation Consulting, at CA Technologies, for contributing to the development of this case study.

J. Orvos, *Achieving Business Agility*, https://doi.org/10.1007/978-1-4842-3855-4_5

Introducing the Solar Corona Insurance Company

Solar Corona is a fictional company that represents a synthesis of various companies that CA Technologies has conducted business with, but the events described represent depictions of actual challenges faced by those companies and CA Technologies in assisting them.

Humble Beginnings

Solar Corona began back in the early 1900s as Solar Corona Mutual, offering both livestock and crop insurance. The company was structured as a mutual in order to provide better benefits to its customers, in that any "profits" made by the mutual went directly back to the policyholders.

Some of the driving forces behind the creation of Solar Corona Mutual were the problems that ranchers and farmers were having with animal illnesses, such as the extensive foot-and-mouth disease epidemics of 1924 and 1929, along with crop destruction through various blights and insect attacks, such as tomato blight and the July 1931 grasshopper swarm that destroyed millions of acres. Crops were also affected immensely by the dust storms of 1934. These conditions, in addition to the creation of the Federal Crop Insurance Corporation in 1938, generated significant demand for Solar Corona Mutual products.

The Federal Crop Insurance Corporation was formed to act as an intermediary between farmers and insurance companies, creating specific contract terms that structured the policy to be fair. The policies required that the farmers insure all eligible acreage for a particular crop, thereby eliminating situations where the farmers—having more knowledge of the relative risks for different acreages—only insured the areas of highest risk.

As the number of policies started to grow, Solar Corona Mutual began to use IBM tabulating equipment in the 1940s for billing and accounting. The IBM 80-column punch card, which had been introduced in 1928, became the record-keeping instrument of choice for the company. They also used the IBM Type 405 Alphabetic Accounting Machine, which had been invested in 1934, to tabulate and print information stored on the cards regarding their customer's insurance policies.

Building the Brand

As other smaller local livestock and crop insurers went under after a number of incidents, Solar Corona grew its reputation for being there through the toughest of times. Since it had policyholders throughout the country instead of localized to specific areas, Solar Corona had a much stronger base from which to pay claims when necessary. It was rare that crop failures would occur across all growing regions throughout the country; therefore, it was unlikely for a massive event to push claims to a point that caused failure.

Self-powered combines were developed in the 1930s, but hit peak sales in the 1950s. As farmers increased their use of machinery, Solar Corona expanded its insurance product offerings to cover combines and other types of farm equipment as they became widespread.

It was Solar Corona Mutual's track record of being there for the ranchers and the farmers that solidified its brand name across the country. As other smaller, local insurers fell and/or could not keep up with the new equipment insurance products, Solar Corona's reputation blossomed into that of a company that could be counted on to meet insurance needs now and in the future.

Growth

As the company grew, they expanded their product offerings to include farm homes and their contents, ranch buildings, barns, grain silos, and other structures as well as providing liability protection. Their core business of livestock and crop insurance was severely challenged during the southern corn leaf blight of 1969 and 1970, but the company was able to survive. Yet competition was growing, and Solar Corona knew that it needed to continue to diversify its product offerings.

Solar Corona found itself short on capital, however, which it needed to keep growing through both new product offerings and additional policyholders. So, in the 1980s, Solar Corona went through a demutualization and became a public company known as Solar Corona, Inc.

The equity markets provided the necessary funds for Solar Corona to fuel its growth through the 80s and 90s. A large portion of the influx of equity was used to pay for acquisitions of other insurance companies that already had established products that Solar Corona was looking to build into their portfolio.

The equity was also used to fund the technology that was needed to support the enormous growth in policies, products, and agents. Things had progressed tremendously from the early punch card days of maintaining records. Now, large mainframe computers and their associated peripherals were the mainstay of record-keeping and calculating.

As Solar used more and more independent agents, vast communication networks were established in order to link the agents' offices to the central mainframe at Solar Corona's headquarters. This enabled the agents to eliminate massive amounts of paperwork and electronically enter information on customers and policies, as well as to provide more personalized materials back to customers in a fraction of the time that it had previously taken.

Stuck on Status Quo

As the number of competing insurers grew, maintaining Solar Corona's position became the most important objective. Because of the acquisitions made throughout the years, the company was ripe for rationalizing its operations, workforce, and technologies.

With each acquisition came redundant departments and computer processing. The strategy that Solar Corona adopted was to first consolidate the computing hardware that each acquisition used to run all of its operations. This was fairly straightforward, only becoming a challenge when the acquiring company used completely different and incompatible hardware. Once the software was up and running on Solar Corona's equipment, they were able to eliminate the expense of the acquired company's hardware.

Once the software was moved on to Solar Corona equipment, teams began evaluating the differences in the acquired company's software and the software used by Solar Corona. In some cases, it was fairly quick and painless to consolidate software systems. This led to subsequent layoffs both in the IT department and in the operations areas that used the software. Typically, this would be areas that were responsible for the basic accounting functions, as well as claims processing.

After the more basic functions were consolidated, attention was turned to working on the more customized types of functions, such as marketing and sales, which were more nuanced according to the particular products and policies that were offered. But Solar Corona prided itself as a beacon of operational efficiency and worked diligently to consolidate as much as it possibly could.

The emphasis on this consolidation stifled any real innovation within the company. Solar Corona lost sight of the drive that created it: to serve customers in the best way possible. It was now a machine, focused on consuming acquisitions and cutting expenses. The mission became only one of cutting costs by streamlining operations and reducing headcount, without understanding the eventual consequences that this would cause.

Current Challenges

Solar Corona faces a number of challenges that need to be addressed.

Losing Existing Customers; New Customer Growth Slowing

Solar Corona's brand was built during a very different time in history. Since the base of Solar Corona's business was built on policyholders in the early 1900s, business is contracting through the natural course of policyholders aging. The initial ranchers and farmers that were the core of Solar Corona Mutual have been retiring and farms have been consolidating. With the advent of better farming techniques, hybridization, insecticides, and other factors, the threat of losing entire crops, although still lingering, had greatly dwindled throughout the years. This has pushed the memory of what Solar Corona stood for and accomplished deep into the back of people's minds.

In addition to policyholder attrition through age, there are those families that are also struggling with trying to maintain their farms amid the growing competition and decreasing margins. As the older generations give way to the younger family members, more and more are seeking cheaper alternatives and switching coverage to companies that can provide more tailored insurance packages to meet the complexity of their needs, all at a more affordable rate.

This changed the very basic driver of what customers were looking for when they shopped for insurance. Yes, there were still family members who had recollections of their parents and grandparents relying on Solar Corona's help during times of tragedy, but those memories were fading. These families were no longer willing to pay the premium prices for coverage that Solar Corona charged because the very nature of what was included in the value proposition had changed. People were not seeking the relationship orientation that Solar Corona had been built upon but were now seeking who could provide the lowest price coverage in an ever-increasingly competitive landscape. New farmers and ranchers were getting squeezed and prices for their goods were falling, so the cost of their policy became the number one driver when shopping for insurance.

Time-to-Market Too Slow

Having focused all of its attention on rationalizing its various acquisitions, Solar Corona was by this time poorly positioned to take advantage of technology changes in a way that would be meaningful in creating new business opportunity. Solar Corona was optimized to peak efficiency, but that optimization came at the price of innovation. There had been such a push to "standardize" and "share" processes and technical capabilities that it had become extremely difficult and complex to introduce new products into the marketplace.

During the heyday of its acquisitions, Solar Corona initially was positioned well to innovate, as each individual company that was acquired specialized in certain types of insurance. This specialization of products offered, and customers targeted, served as the mechanism to maintain the right balance between revenue and expense.

But as time passed and margins shrank, Solar Corona sought to reduce its operating expenses. As a consequence of consolidating the disparate companies it acquired, it lost its ability to truly focus on the individual customers' needs. Their marketing became more generic, their processes more cumbersome, and their reputation one of bureaucracy over caring. When ideas did surface to create new products, they now faced endless discussions, debates, and process and technology hurdles to finally make it to market.

New Products Not Selling Well

When Solar Corona did manage to squeeze out a new product, it was either late or missed the mark in terms of meeting the needs of the potential customers. Because of the myriad number of coverage types and infighting over how best to structure bundles, the types of coverage and policy options available became too rigid and uncompetitive with what some of the smaller, more nimble insurance companies could offer.

In addition to inflexible products, the stories of degrading service for processing claims were becoming rampant.

Introducing Linda

Linda is a respected middle manager in a large company. Linda is caring and friendly. Linda doesn't want to rock the boat and create enemies but wants to make a change in the business. Her personality is an amiable type, caring of how others feel in their organization.

Linda worked for her company for five years and was generally well-liked and respected by her peers that knew her, with very few adversaries. She had a good track record of meeting her annual goals and her job was secure for the long run. She has little personal incentive to go outside her comfort zone and to cause trouble but her core value of looking out for the common good kept pulling her toward action. Linda was a good corporate citizen who cared about the company and its success, and annoyed by poor productivity, inefficiency, process ineffectiveness, heavy manual work, and lack of discipline due to unfocused working efforts. All of these inhibited the company's ability to succeed and grow.

However, Linda was a people-pleaser, and well liked, sometimes needing the courage to stand up and say something that would hurt others' feelings. She was trying to please too many people at the same time, making it difficult for Linda to ask others for favors. She recharged by helping other people. She was caring, encouraging, and sharing. Linda's motto was do it in a caring way to help others.

During 2015–16, Linda's insurance company was facing major disruption from smaller, faster competitors entering the market to attract customers with innovative software products. To combat these new threats, Linda was given the opportunity and challenge of her career: to transform her company into an agile business in order to better sense and respond to customers' needs. But that was a daunting task given that her company, Solar Corona Insurance, was a very large, traditional insurance company that was set in their ways and not inclined to change.

With Linda's new role, she was now in charge of transforming how products are delivered. She faced a lot of uncertainty and new ground for her traditional company set in its ways. Linda had an admirable track record of meeting her goals and delivering on time, and she had ambitions to improve things in the company. But she didn't realize what she had gotten herself into, since now she had to count on business executives to buy in and agree to her cause, not to mention how to do it. She realized that if she failed to get them to agree to change, she would fail. But she was willing to take the risk.

Though thrilled by the challenge, she sometimes felt like a bundle of nerves. She would ask herself:

- How will I be able to make this change happen, when the company is so traditional?
- If I need so many business executives to join my cause to make this a success, how will I get them to buy into changing their ways?
- Since I am not an agile expert, how will I be able to lead the cause with confidence?

Linda had a breakthrough story by applying the four communication strategies to engage the business and extend her company's agility practices into the business. In effect, Linda bridged the business and delivery divisions and pulled it off.

Learn from Linda—Crossing the Business Chasm

Convincing executives to change and adopt agile ways of working into the business is not easy. Business leaders, like most people, are generally attached to processes they feel are tried and true. Linda's questions about how she could bridge the divide between delivery and business were a revealing indication that there were no established change management approaches in the agile community to guide her. Her story illustrates how delivery leaders start off facing resistance as business executives are unable, or unwilling, to link delivery successes with business outcomes that matter. Linda's initial experience was no exception. It was as if she were gearing up to speak to residents of Mars about how great things are on the moon. She felt she had no chance of getting her message across.

What I eventually learned from her story was that Linda had a breakthrough when she started talking about change in terms of what was in it for other executives. She discovered that business leaders will respond to an approach aligned with their own way of thinking, using their own business language—not agile process jargon—and highlighting how they could benefit.

Once Linda realized her communication style was misaligned with how business executives think, she made a serious change. She flipped her approach and started talking to the executives in their own business terms, based on their particular mindset. The result was game-changing. From that point forward, Linda could establish trusting relationships and effectively usher the executives toward achieving business agility.

CHAPTER

6

Sounding the Alarm at Solar Corona

Yearly Budget Meeting: Finance Wants to Cut the Spend

It's 10:00 AM Monday morning as the managers shuffle in to the conference room for the yearly budget review meeting. Oscar Dayton, CFO, is nervously pacing at the back of the room. Oscar had arrived earlier than the rest of the executives, trying to give himself just a few more minutes to prepare for what he knew would be a contentious meeting.

Oscar was a tall, thin man in his mid-40s and a staunchly loyal executive that had been at Solar Corona for 17 years. He was well-liked by most of the other executives, who had come to admire his balanced approach toward being fiscally responsible, while at the same time proudly supporting Solar Corona's major initiatives throughout the years.

J. Orvos, *Achieving Business Agility*, https://doi.org/10.1007/978-1-4842-3855-4_6

But Oscar was worried this particular morning about how the other executives would react to his message. As people gathered in the room and greeted each other, their gaze quickly settled on Oscar's sullen face: he was staring silently at some papers on the table, his horn-rimmed glasses resting slightly below the bridge of his nose. The others made it to their seats, all the while never losing sight of Oscar's troubled countenance. No one could recall ever seeing Oscar so distraught.

As Davis McMillan, CEO, walked into the room, everyone was already seated and the silence in the room was almost deafening. "Good morning, everyone," Davis boomed. The executives around the table each responded, "Good morning," but without the enthusiasm that Davis had shown, and certainly in a more subdued tone than usual, as if they were attending a solemn ceremony.

Davis McMillan was a broad-shouldered Iowa farmboy who had worked his way up to becoming one of the most admired insurance executives in the United States. He was a natural people-person, with an outdoorsy look and infectious smile that resonated with everything that was good in anyone he met.

Davis immediately knew something was wrong. He scanned the faces around the table and stopped when he came to Oscar. He could tell immediately that his CFO was not himself, and had, only through his appearance, impacted the rest of the attendees in a dramatic way.

"Oscar – you don't look like yourself this morning," Davis softly ventured. Davis, being an open, honest, "let's-get-it-on-the-table" kind of guy, quickly engaged Oscar to share what was on his mind.

"Oscar – I can see by your expression you've got something you want to share with us all, and it's probably not good news, is it?," Davis asked. Oscar looked up, pushed back his glasses, and began to speak softly: "Everyone – I've got some bad news. Based on our year-end financials, we are going to have to cut budgets for this upcoming fiscal year."

A collective gasp filled the room. No one could recall ever hearing Oscar uttering those words. "Oscar, let's get right to it – what's going on with our numbers?" asked Davis.

Oscar looked around the room slowly, composing himself as best he could for delivering the news. "Well, I think you're all aware that our core customer base has been eroding for some time now," Oscar said matter-of-factly. "We've been aware of that issue for some time, and we have had plans in place for two years now to create additional sources of revenue." Oscar continued: "We've focused on providing new products for nonfarming entities in the towns where we are well-known, and our brand is a household name." "However," said Oscar in a more somber tone, "these new products are not producing the revenue we had anticipated."

Davis McMillan was shocked. He had been optimistic, based on Solar Corona's reputation, that the nonfarm growth strategy they had developed would more than make up for the loss in revenue that they were experiencing from their core property and livestock insurance business for farmers throughout the Midwest. Based on studies that they had completed, new customer projections were strong, while projections for increasing the business done with existing customers were even stronger.

Where did the projections come from? The marketing department, run by Ann Demming, was responsible for the studies and recommendations that had been made to the board of directors over two years ago. The projections suggested that farmers were beginning to realize that they could make money with their farms ***by not farming***. Instead, by either supplementing their farming activities, or converting them outright, farmers could grow businesses that ranged from on-farm distilleries producing small-batch bourbons to building greenhouse additions that specialized in supplying fresh herbs to restaurants and high-end grocery stores.

Ann Demming had joined Solar Corona four years ago, having quickly built a reputation in prior companies as someone who had her "finger on the pulse of Americans." She was young, only in her mid-30s, but had amassed an impressive track record at several food retail organizations throughout the country, and had been credited with creating marketing campaigns that increased new customer growth rates and existing customer expansion significantly.

Ann's budget had been increased each year over the last four years she had been at Solar Corona. Ann had made many presentations to the board of directors and other executives throughout the company, which resulted in her building confidence throughout the organization that her projections were sound and well thought out.

Davis was particularly optimistic regarding Ann's proposals, since they were based upon building even stronger relationships with the local farm owners and their families, which Davis was particularly keen on supporting. He felt an obligation to people like himself and his family—those that were raised working the land and providing for others.

Marketing Fights Back

Ann was the first to respond to Oscar's hint of budget reductions: "Well, I hope you're not considering cutting the money we need to actually help solve the problem," she exclaimed. Ann continued: "We must continue to invest in our various campaigns to help our customers reach their full potential and build the businesses that will be the engines for our growth."

Davis McMillan had hired Ann for two main reasons: first, she had an impressive background in the food industry, and understood the ups and downs of agriculture; but, more importantly to Davis, Ann had spent some time at a nonprofit organization in her past. Davis saw in Ann a deep-seated desire to create the meaningful story and inspirational connection required to be successful in raising money at a nonprofit and building products and services that the public would come to embrace and believe in the company's mission.

Davis at his core was someone who loved the land and the people who worked it, and he took immense pride and satisfaction from helping those people be successful. Davis had grown up on a farm in Iowa and had suffered tragedy when he was a youngster growing up in Fayette County. He was 10 years old when an F-5 tornado 500 yards wide destroyed a number of farms and homes, including his family's. Fortunately for Davis, his father had built an underground tornado shelter and Davis, his mom and dad, and three of his siblings all survived the tornado. But Tom, Davis' youngest brother, was not so fortunate.

Tom had been playing in the fields next to the family's farm house when the tornado hit. Disoriented by the storm, Tom ran the opposite direction away from the storm shelter and could not be found in time. Frantically, everyone in the family searched for Tom until the last possible moment, but to no avail. Davis' father Michael—knowing he had no choice—pulled the rest of the family together and forced them into the shelter. Davis was crushed to lose his little brother.

From that point forward, Davis focused his energy on helping his family and his community to be better prepared to weather these deadly storms. When Davis graduated college, he knew immediately that he wanted to work at the company that had been there to help his family rebuild their lives: Solar Corona Insurance. The company had been in business already for 40-plus years at the time of that devastating tornado that so dramatically affected Davis' life. Even the name had special significance to Davis. He could still recall the bright ring around the sun that was the first light in the sky that he saw as the tornado pulled away from Fayette County that day. As time went by, Davis eventually learned that the founder of Solar Corona, Ted Simmons, had so named the company based on his own experience of living through a tornado just like Davis and wanted to establish a vision that people could hold as the "sun after the storm"—a helping hand that would shine bright even after the darkest of skies.

"Don't worry, Ann," Davis interrupted. "I'm sure Oscar has come up with a well-thought-out plan to get us back on track while still achieving our mission to move forward."

Oscar, who had been staring down at several charts and graphs, looked up. "Davis, I wish I shared your optimism about being able to cut the spend while preserving most of what we are focused on in terms of marketing campaigns." Oscar paused, but then slowly started to lay out the reality for Davis and the rest of the executives.

"We have four distinct marketing campaigns going on right now," Oscar continued. "Our total spend was designed to be capped at 12% of our current revenue stream when we did last year's budget allocations. This was a bit on the high side, but not unreasonable given our goals," said Oscar. "The issue we have is that our revenue has fallen dramatically over the past year, and we need to adjust the budget going forward to compensate for the loss of revenue," Oscar explained. "And, playing it a bit more conservative going forward would also be my recommendation," cautioned Oscar. "I'm recommending that we cut the marketing budget significantly, down to 5% of projected revenue for the next fiscal year," Oscar said firmly.

Ann immediately lost her composure. "Are you delusional, Oscar?" she blurted out. "You want me to grow this business to make up for that precious revenue that you say we've lost this past year, but you want to cut my budget by more than $40 million?," Ann angrily snarled. "Do you realize that it costs over $200,000 a minute for a prime-time commercial slot on television?" Ann continued her tirade: "If we can't get the word out about our new products, we can't get new customers; no new customers means no new revenue. You – **of all people** – should understand the math, Oscar!" Ann turned to Davis now, looking at him directly in the eye and played directly to his heart-strings: "Davis, I know you genuinely want to help people. How can we help them if we can't get our message to them?," her tone now much more subdued and compassionate.

Ann knew how to plead her case to Davis. As part of her interview process, Ann meticulously reviewed as much history as she could find regarding Solar Corona and Davis McMillan. She was keenly aware of the McMillan family loss all those years ago from the tornado of '68. Ann was confident that Davis would side with her if she could plead her case in a way that would remind Davis of what he and his family had come to rely on in their time of need, as well as all of the other families that had been devastated by that tragedy so long ago.

"We owe it to people who have bet it all to work this land," Ann continued. "They are facing a different kind of storm today, but a storm nonetheless; and it is tearing families apart, threatening the very foundation of what it means to be a farming family in this country," Ann said resoundingly. "Their parents, their grandparents and even their great-grandparents have depended on us for years; times are changing for these families and they need our help in different ways now. She paused for a moment, and then, very carefully enunciating each and every word, exclaimed "but they still need our help."

Davis sat quietly, contemplating what Ann had expressed. In his mind, he knew Oscar would not have come to the meeting and suggested the budget cuts if he didn't believe that they had to do it to be fiscally responsible. But in his heart, every word that Ann uttered resonated so deeply with him and his desire to have his company be the "sun shining through" the darkest of storms. It seemed as if he was caught in a trance, fixated on the dilemma just put in front of him, until Ed broke the silence.

Channels Wants the Ball

"You know there's another way to reach people," Ed Boderman said, matter-of-factly. Ed continued, "And it's much more personal than watching commercials on television. We need to digitally connect with people; we're doing it already, enabling people to pay their bills online and now even on their phones. We can leverage the technology investment we've already made and begin to use it to acquire new customers."

Ed Boderman was the newest of all of the executives, having been with Solar Corona for only 18 months. Ed's background was in digital retail, having spent over eight years at three different retail companies that had made significant investments in technology to enable moving away from brick-and-mortar stores to an online retail model.

"I've watched online retail sales grow from 4% to 9% in total in the last seven years," claimed Ed. "I believe that the future of selling insurance will follow the same model," he continued. "It will start off slowly but will skyrocket as technology improves and the next generation of purchasers becomes more familiar and comfortable with it."

"That's not our model," exclaimed Ann. "Our customer base," she continued, "believes in – and quite frankly has placed their trust in – our agents throughout the country, especially in the farming regions where we are so entrenched."

"Well, Ann, our agents aren't getting it done!," Ed retorted. "Take a look at the latest data. Oscar, I'm sure you've got it in that stack of paper you have in front of you. Please – enlighten us all on how well our agents are performing for us."

Oscar shuffled through a stack of paper and eventually found a few pieces that related to agent performance. "I'm sorry Ann," Oscar started speaking softly, "but Ed has a point. Our agent productivity in terms of actual net new sales is down from this time last year by 30%." "In addition," Oscar continued, a bit more forcefully now, "we've been losing agents and we're having a difficult time hiring new ones and getting them up to speed to the point where they are positively contributing to revenue."

Ed couldn't resist grabbing onto Oscar's last statement: "So, Oscar, what you are saying is our agents are not producing the growth through their relationships, and our marketing campaigns – which are costing us millions – are not producing results, either. Is that right?"

Oscar looked up at Ann, sheepishly, and then over to Davis, and replied "Yes – that's correct."

Davis interrupted the exchange: "So, Ed, what's your proposal?"

Ed paused for a moment, and then carefully, intently started to lay his thoughts out on the table: "I know I'm relatively new here. I know I haven't grown up in the insurance business. And – although you may find it hard to believe – I certainly understand and respect the relationships that we have with our customers, largely based upon years of dealing with our agents."

"But times are changing," Ed continued. "*People* are changing. We need to change as well. We need to fully embrace technology and what it can do for us that our agents and our television ads simply can't do. When we run our television ads at $200,000 a minute, what kind of return are we expecting on that investment?" Ed asked, rhetorically. Ed knew Solar Corona was not getting anywhere near the return Ann had alluded to in her prior proposals that were sold to Davis. He continued: "And as far as our agents go, how much does it cost us to hire, train and pay an agent, just to have him or her turn around and leave?" "Oscar," Ed focused his attention directly on the CFO now, "you claim we need to cut the budget. Well, I don't think not spending the money will solve our problems. What we need to do is get more return on the money we are spending," Ed exclaimed. "If we started looking at these 'expenses' as money we are 'investing' in Solar Corona, are we making wise choices? I think not."

Ed had the room for the moment, as everyone was interested to see where he was going to take this line of thought. Ed stood up now and walked around the large oak conference table as he spoke: "We need to invest our money in a way that provides a much better return on the investment we are making. We need to be taking precise, targeted actions instead of blaring commercial ads on television to millions on the hopes of getting a few. We need to be lowering the cost of educating our potential and existing customers about our new products, and enabling them to service themselves, instead of paying a high cost for agents that can only handle so many interactions a day."

Davis was intently listening to Ed during his entire speech. He was watching the faces of the other executives: he saw the anger and resentment in Ann. "I know what Ann's thinking," he said to himself. "This company was built on relationships; relationships that were forged between our agents and hard-working people that Solar Corona had protected for generations." Davis shared her sentiment as well, but he also knew his CFO and trusted him implicitly. If there was one thing that was more terrifying to Davis than potentially damaging the very nature of the relationship the company had with its customers all these years, it was potentially losing the company altogether—not being there for the millions of people that relied on them. Davis decided right then and there that the latter was not an option. He knew he needed to do something. But how much of a risk was Ed asking him to take?

"Ed," Davis spoke as he leaned forward and looked directly at him, "I'm intrigued by what you are suggesting; however, we simply can't 'bet the ranch' on a notion that may or may not solve our problem. I know you've seen this done in the retail space, but if I recall, none of those retailers closed up their shops all together, at least not until they were comfortable that they had a technology solution that would generate a consistent, significant revenue stream."

"You're right, Davis," Ed replied. "They kept their brick-and-mortar locations open while they built their online retail presence." Ed continued: "There's no reason why we can't continue to use agents, and maintain most of our branches, while we invest in the technology to compete online."

Davis turned to Linda Grable, Solar Corona's CIO, and said "Linda, what are your thoughts on all of this so far?"

Delivery Is in Shock

Linda Grable was the most senior person in the room, having been with Solar Corona for 23 years. She had risen through the ranks, having joined Solar Corona as a technology unit manager in the mid 90s. She had grown up in the information technology arena, having gotten her degree in computer science and having worked the early part of her career as a programmer for a regional bank in the Midwest.

"Davis, I think there could certainly be something to Ed's plan," Linda began to respond slowly. "But – as Oscar has pointed out – we're not achieving our current revenue targets and boosting our IT capabilities to be able to undertake the plan he is proposing would mean a major investment in people and technology."

Linda's head was still spinning. She was recalling all of the effort that they recently went through in IT to be able to provide additional functionality for the call center, as well as for claims processing. There had been some preliminary discussion about providing the capability for people to pay their bills and submit claims via their smartphones, but what Ed was describing was way beyond that; Ed was venturing into presales and policy purchasing, which was not something that had been discussed in the past in terms of putting it the hands of the customer as opposed to solely controlled through agents.

"Linda? … Linda? …" There was a voice that snapped Linda out of her momentary lack of focus. It was Davis. Linda replied "Yes?" Davis, having gotten Linda's attention, started to ask questions regarding her response: "Linda – I agree we will need to invest, but I'm not even at that point yet," exclaimed Davis. "I'm still not convinced that a) we know enough as a company to make this happen, b) that our customers would use it, and c) that it will actually solve our revenue issues." Davis paused, and then added: "And I have no idea how to do this in a way so as to not send a negative message to all of our agents and have them all get up and quit."

Linda almost felt a bit embarrassed, having immediately focused on what it would mean for the IT area and not thinking more broadly, as was Davis. How could she have been so selfish, she thought to herself? Linda began to quickly consider the issues raised by Davis. She sat back in her chair and pondered to herself, "I don't have the answer to those questions right now. How can I answer those questions in a way that will help Davis make a decision about the budgets today?"

Linda's mind swirled for a bit and then she remembered something. They had recently been using a new approach to develop some of the new software that would replace some of the old, outdated systems that they had been running on for years. This new approach—known as agile—offered a new style of working and producing software that had started to produce some positive results. The piece that struck Linda as possibly being useful was the way in which they ran these cycles called "sprints." Since the outdated systems that they were replacing were so old, there weren't a lot of people in the organization that really understood how they worked. One of Linda's teams had learned about using agile and working directly with the business to deliver software in short cycles, put it in production, and get immediate feedback so as to steer continued development. This meant that they learned together, took some level of risk, but limited the exposure to only two weeks' worth of work if they got it wrong. This way, with the fast feedback they were getting from people who used the system, they could adjust for the next sprint and try again.

Linda broke her silence and looked at Davis. "I have an idea," she said. "I think that maybe we can apply some of the agile techniques that we have been trying out in IT to work in sprints to test out Ed's proposal."

"Agile techniques?," asked Davis. "Sprints?," asked Oscar. Oscar continued, not realizing that he had interrupted Davis: "What's a sprint? You folks in IT get together and run around the building or something?" Oscar chuckled, followed by the rest of the room. It was the first release of tension that they had experienced since the start of the meeting. But things quickly became tense again.

Linda began to explain what the teams had been doing. She started discussing all of the various practices and metrics that they were employing as part of their agile journey. But it was falling on deaf ears.

Davis muttered, "Linda, you know I respect your technical leadership here at Solar Corona. But, honestly, I have no idea how what you just described is going to help us answer the questions I asked earlier." Oscar chimed in: "And from where I'm sitting, I didn't hear anything that gives me a comfortable feeling that we'll get our spending under control."

Linda looked at everyone in the room and realized what she had done. It was as if she started speaking in an entirely different language to her colleagues. Linda quickly adjusted: "Davis, and everyone else, I apologize. In my desire to provide an answer, I started describing something that has no meaning to any of you – yet." Before anyone responded, Linda continued: "However, you all know me, and my reputation speaks for itself. It needs to be explored, but I think I might have a solution; what I need from all of you is a leap of faith – at least for a little bit – to allow me some time to come back to this group with a proposal in language that you can all understand."

The group agreed to give Linda one month to come back to them with an explanation they could all understand.

Delivery Takes the Lead

Linda left the meeting feeling both exhilarated and frightened. She was excited about the prospect of potentially using agile to solve the problems that Davis had outlined in the meeting. At the same time, she was at a loss for how she would find a way to speak a language that the business understood. The day after the meeting, Linda invited a team that had been working on utilizing agile on the new systems replacement to join her in a discussion around Davis' questions.

The team was comprised of several members of the IT organization and several members of the business community that were heavily reliant on the software that they had been replacing. Linda began to address the room: "I want to thank all of you for coming to this meeting on such short notice. But I have a challenge and I need all of your input to help try and resolve it. Yesterday, I was in a budget meeting with the other executives and Oscar, our CFO, suggested that we need to cut budgets." There was a collective gasp in the room as members of the team immediately became concerned that the budget for their project would be cut. "I assure you," said Linda, "we did not talk about your particular budget." She then began to explain the history behind the budget cut proposal, specifically as it related to Solar Corona's inability to keep its costs down and its revenues growing.

Linda started to explain in more detail: "As you all know, Solar Corona has been challenged over the last several years to increase its revenue and decrease its costs. Quite frankly, this is one of the reasons why we're engaged in the project that you are all working on, since the existing system that you're replacing is so inefficient and requires so much effort to run and maintain. One of the reasons that we selected agile to use as the methodology for this project is because it would enable us to get functionality delivered much more quickly, targeted to specific high-cost areas that we could focus on and begin to take out some of the cost as we go."

Susan, who was the scrum master for the team, responded: "So, Linda, are you asking us to expand the team and the scope of what the project is working on in order to replace even more systems so that we can cut even more cost?" Linda thought carefully, and then replied to Susan: "I'm encouraged by the progress that this team is making but I'm not sure how we can apply it on a broader scale."

John, a development team member, asked Linda: "What is it that you're concerned about?" Linda replied, "Well, John, when I was in the meeting with the other executives and I mentioned sprints as an example, none of them had any clue what I was talking about." Susan, with her scrum master background kicking in, immediately responded, "Linda, that's simple: what we need to do is train all of the executives about agile."

Linda looked around the table, and asked the group a rhetorical question: "How would you feel if you had to learn something about finance or marketing and use their language? The members of the team looked at each other and began to realize what Linda's point was: she was highlighting the fact that the language around agile meant something to the people on the team who had gone through training and had been living scrum day in and day out. But as an organization, we haven't trained our leadership—within either the business or the delivery organization—on what scrum is, and there's not enough time to learn it in order to apply it on a broader scale.

George, the product owner for the team, chimed in: "Yes, I can say unequivocally that it was a challenge for me, coming from the business, to learn all of the jargon that exists around agile. But once I learned how I could apply it toward what we were trying to accomplish from a business perspective, I was much more comfortable. What was meaningful for me was to learn not just what the words meant, but more importantly what the intentions enabled me to do as a product owner in achieving the business results that we're looking for."

Linda looked at George emphatically, her voice conveying her excitement: "Yes! That's exactly what went through my mind when Davis presented his challenge to the executives in the budget meeting. I thought to myself that if there's a way we can enable the executives to understand the ***business impact*** that agile has, they could begin to see how to apply it toward solving the larger problems that we have here at Solar Corona."

Delivery Sees the Value—But the Business Is Not Convinced

Emily Rogers, one of the key managers in claims that was part of the small group that Linda had brought together, jumped into the conversation: "George, you always were the optimistic one! It's fine to apply these techniques that we are using on this isolated effort right now, but I can't even imagine how we could scale this in terms of getting business involvement where it's needed." Emily continued, "We don't have a pool of people sitting around on their hands doing nothing. Every single one of us in the business that you folks in delivery need to provide input and guidance on all of these 'stories' that you have us working on is putting in overtime. We have to find ways of getting our normal work done, while at the same time supporting these efforts. How would it even be possible to expand this to the whole company?"

Before George could even reply, Linda started to speak: "Emily, I know you and a lot of others have been burning the candle at both ends and are getting tired. I'm not suggesting that we continue with things in the same way that we are currently. It's not sustainable. We've been working this way to try and jumpstart things, but we need to do better if we are going to apply what we are doing on a broader scale."

Emily responded, "You're damn right, Linda! You folks in delivery have been pushing this new way of working, but the higher-ups in the organization don't know what's going on and middle management is afraid to speak up. If this is going to scale, we need everybody on board."

"You are absolutely right, Emily," Linda said. "And that's why I have brought everyone together here so that we can all agree on how to accomplish getting them there. You see, I suggested applying agile to help solve these problems in our executive budget meeting with Davis the other day, and he looked at me as if I had two heads." Emily chuckled a bit, which helped release some of the tension that had been building in the room. Linda continued, "We can't educate everyone at once and magically have the entire organization agree to work this way. We need to slowly get their buy-in by getting them not to understand agile, but first to understand why they even need to listen to us."

Emily responded quizzically: "How on God's green earth do you expect to do that?" Linda looked at her and smiled. "Emily, how did we get you to buy-in to work this way?," she asked. "Well, to be honest, I looked at it selfishly. I realized that if we replaced that old rust-bucket of a claims system with something new and shiny, then I would have a much easier time doing my job!," Emily responded. "Exactly!" Linda replied, "You found what was in it for you, and that's what made you want to see it succeed!" Emily nodded in agreement as Linda continued, "You see, not everyone wants to make a change to the way they work, unless there is something in it for them. We need to help them see how each and every one of them is impacted by what's happening out there."

Developing Impact Statements

Linda now had everyone's attention in the room. What she was saying made sense to them. Everyone has some sort of "pain" when it comes to the work that they do, and if there was a way to help them see that pain getting resolved by working in a new way, they would—at least potentially—listen.

Linda started by describing how she saw approaching finance. "You see," Linda said, "Finance has a very narrow view of the world, especially now that our revenue is down. What they see is the gap between revenue and expense getting smaller and smaller, and their only conclusion is to cut expense. So, we need to help finance see that there are other ways to widen the gap. Our competitors have financed and launched products that, because they are priced higher, bring in more revenue. We, on the other hand, have been investing in products that have become commodities; our prices have been cut to the bone. Our return is minimal."

George jumped back into the conversation. "Linda, in my product owner role, I've been researching our competition in order to help me understand what we're up against. Maybe I can pull some spreadsheets together that can tell the story around what our competitors are investing in and how much return they are seeing." Linda replied, "That's great, George; you know we'll need that kind of detail for Oscar!" Everybody in the room laughed.

"Now, anyone have any thoughts about marketing?," asked Linda. John from development piped up: "Linda – I think I may have an idea, but it's going to need your approval." Linda replied "OK, John – I'm certainly open to suggestions. Let's hear your idea."

"Well, I'm friends with two of the developers – Rick and Mary – on the data warehouse team," John began. "And we get together for lunch often and invariably talk about the projects we work on and people we deal with. The other day at lunch we were talking about all of the different marketing projects and how they were very frustrated." Linda asked, "What was it that they were frustrated about, John?" John replied "Well, it seems that we have all of these different marketing campaigns that are aligned to the various insurance products that we offer. In speaking with Rick and Mary, they see a very siloed approach toward the way the campaigns are analyzed." Linda pressed John to continue.

"The problem that Rick and Mary see is that – because the campaigns are siloed – we get analytics on each campaign, but we're not getting analytics on the people." replied John. He continued his explanation: "Because the various product groups are competing with one another, they don't readily share information. We could be getting so much more information about the person, but we're missing the opportunity. In addition, we're not aggregating the data across campaign methods – you know, e-mail versus television as an example."

Linda paused for a moment and then replayed what John said back to the group as a whole: "So, John, what you're saying is that Rick and Mary are frustrated because we could be learning so much more about our potential and existing customers if we approached them in a way that was cross-product and using cross-marketing vehicles, is that right?" "Exactly," John replied. "We should be positioned to be able to know how an individual or a family is responding to our products and through what mechanisms and why so that we can focus our interactions with them. This way, they get the information they need, and we get more value from our advertising by being focused."

"So," Linda replied, "what is it that you need from me?" John replied, "I have an idea on how we can combine some of the data in the various campaign warehouses so that we can get more precise analytics on where we get better conversion rates. But I need some time from the data warehouse team to pull that data together." Linda looked at John and replied, "You got it – make it happen."

Next, Linda knew she had a very different sort of challenge with Ed in channels. Unlike most of the other executives, whom Linda needed to prod along to see how things were changing, Ed was different. He had recently joined the company, with experience in online retail, and was gung-ho in trying to quickly get his strategy of direct digital sales in place. But Linda knew that both agents and direct digital was going to be necessary for Solar Corona to get back on track.

Linda, looking around the room, searched for someone that she knew could help her. Her eyes circled the room and stopped about three-quarters of the way around when they met up with Alice Gray. Alice had been involved in several of the projects that were Solar Corona's initial forays into direct digital services, with some very simple applications features that enabled customers to see their bill and pay it online. Linda remembered that Alice was on the "front lines" when the company did the initial research on starting to provide the service, and also that Alice had handled some of the feedback from some different customer focus groups that they had run early on to get feedback.

"Alice," Linda began, "You and I both know Ed and the challenge we're going to have with him, yes?" Alice nodded in agreement, "Yes, Linda, I'm well aware of Ed's background and what he would like to do. We need to give him a new pair of glasses that he can look at things through – is that what you're thinking?" Linda smiled. She knew Alice was in tune with what she had in mind. "Alice – you read my mind. Do you think we can set up some visits?," Linda asked. Alice smiled back and said "Sure – I think I know exactly who Ed needs to meet."

Now all Linda needed to do was to get the various leads in her delivery organization to come together.

Learn from Linda—Sound the Alarm

Facing this daunting task, Linda was nervous because she knew the only way to succeed was by convincing executives in her company to do things differently. Simply falling in line with the current operating model would likely mean failure, so she resolved to push forward. She was not exactly sure how she would do it, but she knew one thing: it was solely up to her to get all her business executives, not just delivery, to embrace agile ways of working across the company.

Linda would arrive at the office with the trepidation of knowing that she was about to contact busy executives to ask them to put more work on their plates and implement agile ways of working into their departments. She made many attempts to engage the business executives with no success.

When she thought how improve her situation, Linda decided to contact the executives with a different approach. Rather than asking them to change to agile, she needed to alert them to an emerging problem that will impact their success. Preparation for this approach would entail research into a looming threat for the business, which would enable her to speak knowingly of and hypothesize about a troubling situation, providing advance notification to the executive.

Linda took this approach and found a distinctly different result on her next call. She can still remember the VP's voice on the phone. There was a pause and then his tone shifted. He was intrigued and wanted to learn more. He was not just appeasing Linda; he was paying attention and expressing genuine interest in meeting with Linda. Why? Because the conversation she had initiated was focused on something he cared about.

From that point on, Linda actually enjoyed asking for executive meetings. She presented her case with a palpable sense of purpose since she had something meaningful to share with the business leaders. As Linda put it, "I knew I could give them a real reason to meet with me. I was able to sound the alarm, making each executive aware of a particular issue that threatened to impact them negatively."

CHAPTER

7

Solar Corona Looks in the Mirror

Stopping the Dance: Delivery Recognizes the Dance

Linda knew that it was up to her to "stop the dance" that was constantly in play between IT and the business. Linda, who had studied organizational leadership as part of her education, remembered a book she had read years ago called *Seeing Systems*[1]. In the book, the author described the relationships among those at the top of an organization, those at the bottom, those in the middle, and finally the customer. Linda realized that the executives (the tops)—including herself—were constantly faced with complexity and uncertainty and rarely established the right kind of relationships between the middle managers (the middles) and the teams and individuals on the ground "in the trenches" (the bottoms). This led to the executives taking on the burden of more and more responsibility, while the bottom relinquished responsibility at the cost of constantly being told what to do.

[1]Barry Oshry, *Seeing Systems* (San Francisco: Berrett-Koehler Publishers, Inc., 2007).

J. Orvos, *Achieving Business Agility*, https://doi.org/10.1007/978-1-4842-3855-4_7

Linda, realizing that this was a large part of why Solar Corona was in trouble, was quick to agree to a teams' request to adopt agile methods for the system replacement project that was currently underway. She had realized that agile was a way of "stopping the dance" among the tops, middles, and bottoms. Agile removes the connotation of a hierarchy among the members of the team, and instead replaces it with a team full of equal members. The team then has to choose how it will function as a team to fulfill all of the responsibilities that went along with top, middle, and bottom roles. The challenge that Linda faced now would be to get each of the executives (tops) to relinquish a degree of control and create teams instead of hierarchies.

The impact statements that her team developed would be the catalyst to help break away from the continuous spiral of blame that had been plaguing Solar Corona for years. Using the impact statements, and meeting one-on-one with the executives, Linda would try her best to stop the dance.

Linda decided to meet with finance first. Her reasoning was that she needed to get Oscar on board with her plan, otherwise there would be nobody to supply the necessary funding in a way that was commensurate with using the agile approach. The challenge would be to raise Oscar's awareness level enough about what was happening and then to generate enough of a desire within him to change the way the finance department funded things.

Finance Looks in the Mirror

Linda purposely scheduled the meeting at a convenient time for Oscar and suggested that they meet in Oscar's office so that he would feel comfortable. When Linda showed up the day of the meeting, Oscar was ready and waiting for her, with a lot of spreadsheets and graphics to explain his position on the budget.

Linda sat calmly, while Oscar outlined his position with regard to how poorly they were doing financially. Linda was aware of this already based on the work that she had done on crafting the impact statement, but she let Oscar continue in order to build up a sense of trust and partnership with him. Linda listened intently as Oscar went through his spreadsheets and graphs.

"Oscar, thank you for sharing all of this wonderful information with me," said Linda as Oscar completed his presentation to her. She continued: "You certainly have a tremendous grasp on what's happening with Solar Corona's finances. I think you're able to tell a good story around our declining revenues and why you feel it's necessary to cut expenses proportionally. However, what I'd like to do is offer an alternative to cutting the budget that may actually help us to generate revenue and become more competitive at the same time."

Oscar seemed intrigued by Linda's proposal."OK," he said."I'm willing to hear your proposal, but I hope you brought objective data for us to review that backs up your position."

This was the opportunity Linda had been hoping for; Oscar had opened the door and now all she needed to do was walk through it.

Linda slowly began to review the impact statement that her team had crafted so carefully. "Oscar," Linda exclaimed, "I believe our problem is not about the amount of money that we're spending; it's the amount of *revenue* that we are getting for the money that we are spending. Let me show you some data around our top five competitors' spend." Linda pulled out the spreadsheets that her team had put together to share with Oscar."You see, Oscar, we have been investing in products that don't provide any significant returns. However, our competitors are spending more money than we are and at the same time they are seeing three times the return on investment than Solar Corona."

Oscar seemed puzzled."Linda, why do you think our competitors' revenues are so high?" asked Oscar. Linda replied: "Well, I believe there are two main factors that contribute to this statistic. First of all, our competitors only fund projects incrementally. This means that instead of a project getting the full budget for the fiscal year, the IT organization only funds work on a quarter-by-quarter basis, focusing on how much value that the software they delivered provides to their customers."

Oscar was astonished. He looked at Linda quizzically and asked, "Are you actually saying that you're willing to tie project funding to how much value the software you're delivering is providing?" Linda replied: "Yes, Oscar. I've spoken with the IT teams and they agree that we're all in this together. If we can control the budget to ensure that we get an appropriate return on our investment, would that be meaningful to you?"

Oscar replied: "Yes, having control so that we can adjust budgets to the places where we are having the most success would be a great benefit in controlling expense." But Oscar was puzzled."Linda," he asked,"why haven't we adopted this already?"

Linda started to explain: "Well, Oscar, the way the funding works today – which you know as well as anyone – is that each business unit funds the work that it wants to do. There's very little oversight as to how much return on investment you're expecting from each of the investments that we make in IT." Oscar was startled and quite frankly a bit upset by her statement."Linda, are you saying that I'm not doing my job?," Oscar retorted. Linda quickly reassured Oscar that was not what she was implying. What she was describing to him was the fact that each business unit had its own budget and depending on how it viewed investment there was really no test on return until it was too late in the game.

Linda highlighted that all of their measurements were lagging indicators and they did predict whether or not the work that they were doing was actually going to generate the revenue or reduce the costs as predicted. In other words, there were no measures to let them know whether or not they were heading in the right direction. Oscar was beginning to see Linda's point. He replied: "So, what you're telling me is that these other companies that are getting higher revenues are doing so because their expense is focused on areas where the return is the greatest?" "Exactly," Linda said; "they focus on only the work that's going to ensure that they get a good return on investment."

"Well, what happens if they're not getting a good return for their investment?," asked Oscar. Linda replied: "That brings me to the second point of what they do differently. These companies are willing to stop projects that are not generating a healthy return. Then they evaluate how the money could be spent differently to generate a higher return on investment." Oscar said, "Linda, that makes total sense, but I think we're going to have a problem. All of the business units are used to getting a yearly allocation and then spending it in a way that they think will generate good returns. Here in finance, we have very little say once the budget is released."

Linda acknowledged his concern. "Oscar," she replied, "that's why I'm coming to you first. I think that in order for us as an organization to better control revenue and expense we need to change the dynamic of how we provide funding to the different business units and how we govern them within the course of the year." Oscar sat back in his chair and pondered. He slowly sat back up, placed his arms on his desk, and looked at Linda and said, "I get it. We need to make the executives within each business unit responsible for the return on investment that they are predicting. How can we make this happen? How do we make such a drastic change?"

Linda looked directly at Oscar and replied, "Well, Oscar, I think you're right – it is a drastic change. And we can't change the entire company overnight. There are too many people and too many business units. In addition, we've been doing it this way for so long that people will be skeptical that it will be successful. We will need to find a unit where there's someone in the business to support working this way. We do the best that we can and steer funding toward investments that will make a difference, and then show the results to the rest of the organization."

Oscar smiled and said, "I think I know who we can get on board; let me get back to you tomorrow." Linda left Oscar's office, cautiously optimistic about what had just happened.

Marketing Looks in the Mirror

Now that Linda had set the stage with finance, she set her sights on marketing. Linda knew this was probably going to be one of the hardest discussions to have, since Ann was so negative in the budget meeting the other day. But Linda felt confident that the impact statements that she and her team crafted for Ann would be enough of a positive message that it would persuade Ann to join in and try the new approach.

Linda was anxious, sitting in Ann's office waiting for her to return from another meeting. It had been two weeks since Linda had requested time with Ann. Linda didn't know if Ann was purposely putting her off or if she was indeed that busy. She tried not to think negatively and focused on the outcome that she was aiming to get out of the meeting today.

Ann walked into the office 20 minutes past the meeting start time and barely acknowledged Linda sitting there at Ann's working table off to one side of the room. Linda stood up and greeted Ann with a smile, even though she was a bit put-off by Ann's lack of respect for her time. "Thanks for taking the meeting, Ann," Linda said as she shook Ann's hand. "I know you're extremely busy and I appreciate you taking the time to discuss some ideas I have with respect to our budget meeting and Davis' request."

"Well, I guess I don't have much of a choice," Ann scowled as she replied to Linda. "It seems you're Davis' chosen one around here. He clearly believes that you can solve his budget problem." Linda immediately sensed the hostility that Ann was harboring toward her. She could tell that Ann was highly upset and it would take some effort to get her to open up and have an objective discussion.

"Ann, I can see you're upset about this whole budget topic," replied Linda. "But I'm not immune to budget cuts, either." Linda continued: "I've had my fair share of cuts to 'take one for the team' over the years and have been pressed to still get all the work done. I can't tell you how many times I've heard the phrase 'do more with less.' And when we operate that way, we all suffer the consequences: overtime, lack of focus because too many projects are competing for resources, and unhappy customers." Linda finished her rebuttal by saying, "Ann – I've been struggling with ways to cut spend in IT, so I feel your pain. But, I think there is a way that Solar Corona can solve the problem."

Ann slumped down into a chair at the table, visibly exhausted. Linda knew that she needed to have the discussion in a different setting, to get Ann out of the office and feeling more upbeat. "Ann, I tell you what. I'll buy you lunch today," exclaimed Linda, in an almost excited tone, as if inviting someone to a party. "Let's get out of the office and take a break. If we feel up to discussing the budget, fine. But if we don't, that's fine as well. I just want us to be able to get away from the stress of being here for a couple of hours." Linda was looking at Ann's expression, which was completely blank. She couldn't tell if she was going to accept or explode right there in the office. A few seconds passed, but it seemed like an eternity to Linda, when finally Ann replied: "Sure, let's get out of here for a while. I can certainly use the break!"

Linda and Ann went a few blocks north and east of the building, to a small seafood place called "Captain Jack's." Linda and Ann went to a quiet corner, so they could be able to hear each other. After ordering, Ann looked at Linda and said "Gee – I didn't realize how long it's been since I actually had lunch away from my desk." Linda smiled and said, "Me too, Ann. We need to do this more often!"

Ann replied, "Yes, but we need to do it under different circumstances. This whole budget thing has really taken its toll on me!" Linda smiled and said, "Ann, may I ask you a question?" The waiter was now at the table, bringing their lunch orders over from the kitchen. Ann did not reply immediately but waited for the waiter to finish bringing their orders. As the waiter left, Ann replied "I suppose." Linda paused to give Ann time to try her food to make sure she was happy with her order. Linda slowly got up the courage to ask the question: "Ann, I know the television ads we run are quite expensive, right?" Ann looked at Linda, not knowing where this was going. Ann replied, "Yes – very expensive. Prime time slots are the worst, but even off-peak hours are still pretty pricey. Why do you ask?"

Linda paused for a moment and then slowly began to reply. "Well, I was just wondering: how do we know who we're reaching out to with those ads? I mean, we don't know who watches those shows that we're buying ad time in, do we? And if we really don't know who's watching, how can we be sure that the message is getting heard by the right people?"

Ann seemed annoyed at first, but the relaxed atmosphere in the restaurant, along with a terrific lunch order of sea bass, seemed to calm her down. Ann explained things to Linda: "The networks have fantastic data on the demographics of each show's audience. We utilize that data to understand what shows and times we should run our ads."

Linda thought for a moment, and continued, slowly exposing more of the data that the team had crafted in the impact statement for marketing. "Well I guess that's pretty good information and helps Solar Corona in getting a general understanding of the potential audience for our commercials, but how do we know if the people that saw the message become actual customers of Solar Corona?," Linda asked. Ann replied, "Oh, that's easy. The network provides us with data on the average population size that watches the show and therefore has potentially seen our ad. In our ad, we provide a code for people to use when they either call or log in and ask for information. So, we know exactly where our leads come from." Ann sat back in her chair, seemingly confident in her answer to Linda. Linda, already knowing the answer, continued to press

Ann. "So, you've got the size of the total population, as well as the data on who actually purchased?," Linda asked. "That's right," Ann replied. And then Linda dropped the big question: "So, are our television ads more or less cost effective than other methods?"

"Well, I have to admit that lately our television ad conversion rate has been fairly low," replied Ann. This was Linda's chance: "So, Ann, if it was your money that Solar Corona was using to advertise, would you continue to spend the same amount on television ads?" Linda realized this was dangerous territory to get into, but she felt she needed to connect Ann to the spend to engage her in a way where it felt more personal.

Ann thought for a moment and replied: "I think I would want to get a better conversion rate out of the investment. The problem is we don't know who is watching and therefore we don't tailor our ads; moreover, we purposely leave them a bit more generic, with the hope that it will appeal to a wider audience." Linda grabbed the opportunity that Ann just provided to her. "Well, what if we could increase our conversion rate to be 5 or 10 times what it is today?," asked Linda. "Would that be a win for Solar Corona?"

"That would be fantastic," exclaimed Ann. "And how do we do that?"

Linda replied, "That's the next step, Ann. I've got some ideas, but I want to share them with the entire team. Can you wait a few more days?"

Ann, as she finished her dessert, replied "Absolutely – and next time, lunch is on me!"

Channels Looks in the Mirror

Now that Linda had Oscar and Ann's attention, she knew her next challenge would be talking with Ed. Linda knew exactly how to approach Oscar and Ann, but she was a bit bewildered as to how to speak with Ed. She knew Ed would be excited about driving more marketing through the digital channels that he so desperately wanted. Ed had deep experience with digital channels from his time spent working in retail. This was not what Linda was worried about. Her main concern was how to fold the digital channels Ed wanted with the other channels, primarily the agents. This is what Linda was counting on Alice to arrange.

Linda knew from her experience in banking that automation can play a large role in helping offload expense by enabling people to become their own banker and execute most transactions online. However, she was also well aware that for certain transactions customers desired, and in some cases demanded, access to people to either help them to some extent, or to fully perform the necessary activities required by certain products and services. In addition, she knew that customers can become very demanding, especially when something goes wrong and they need help. Linda remembered many challenges they had in unifying information and access to the customer, the call center, and the various specialties, like mortgage processing, so that customers didn't get frustrated with having to supply information over and over because the digital channel was not interwoven with the other parts of the bank. This was the challenge she knew had to be addressed in her conversation with Ed.

Linda drove up to Ed's house, right on time to pick him up for their field trip that Alice had arranged. Linda had persuaded Ed to join her in visiting two of Solar Corona's customers that day to hear first-hand their feedback and desires for an improved service model. Linda was sure that if she could get Ed to understand that it wasn't about choosing digital over the agents, that it was really about enabling both channels and focusing them on the customers who were best suited for each, then Solar Corona would have created a win-win scenario.

Ed was now at the car door. Linda unlocked the door and Ed got in the front seat next to her. "So, where are we off to?," asked Ed. Linda replied, "Well our first stop is going to be with Mark and Gail Baker, fairly new customers of ours. Both sets of parents are customers of ours as well, going back now for over 30 years. Mark and Gail were married just a few years ago and have opened up a family-run small business making small-batch bourbon with the corn their parents' farms produce, either in times of low demand and prices have fallen or when the harvest produces excess yield because of good growing conditions."

Linda and Ed pulled up to Mark and Gail's distillery. They went inside and were greeted by Mark, while Gail was presiding over a tour of the distillery for a group of bourbon enthusiasts. Marl led Linda and Ed into the back of the distillery and up a set of stairs to a small, glass-enclosed room that was primarily used to oversee the distillation process. Linda, Ed, and Mark all took a seat around a large oak table.

Linda began the conversation: "Well, Mark, first I want to thank you for seeing us today. We know your time is valuable, and we sincerely appreciate you making some of it available for us to discuss how Solar Corona can better serve you and your business. Let me introduce my colleague, Ed Boderman. Ed is a relatively new addition on our staff and has tremendous experience in 'going digital' with respect to how we interact with customers like yourself."

"Pleased to meet you, Mark," responded Ed. "I'm very excited to learn all I can with respect to how Solar Corona can better serve you, especially in the new world of digital." Mark replied, "I'm pleased to meet both of you, as well. My parents have been customers for over 30 years and that's what drove us to become customers as well. It's great to see Solar Corona being more proactive with regard to understanding our small business insurance needs. Frankly, I've been getting a bit frustrated with the lack of digital access you provide in order for me to continually monitor my insurance needs, update my policies, and take care of claims."

For the next two hours, Ed, Linda, and Mark discussed the types of services Mark and Gail's small-batch bourbon business would be looking for as they grew, and how they envisioned being able to do what they needed to do without having to wait for an agent to come see them. Ed was very excited; he felt he was hearing from an actual customer as to how they wanted to interact with Solar Corona, and it was everything that Ed had been pushing for since he joined.

Back in the car, Ed and Linda chatted while they drove to the next customer on their itinerary. Ed was so pumped up he could hardly contain himself. He was still all smiles as they pulled up onto a long driveway winding along up to a farm house for their next visit. Ed asked Linda, "So, who are we seeing now, Linda?" Linda replied: "Ben and Nancy Baker. Mark's parents." She said this with a wry smile coming to her lips. Ed said "Great – Mark had so many great ideas, I'm looking forward to meeting his parents to hear even more!"

Ben and Nancy greeted Linda and Ed at the front door to their farmhouse. Ben was tall, in his sixties but in terrific shape, with white hair and a typical farmer's tan that left his forearms bronzed, but alabaster white where his shirt sleeves began. Nancy was a bit shorter than Ben, but with a personality that energized the room. You could see the years in her eyes, though, as she greeted Linda and Ed and invited them inside.

Again, Linda began the conversation, just as she did at Mark and Gail's distillery. She explained why they were there, but this time Linda and Ed got a very different reaction from Ben and Nancy. Just a few minutes into the discussion, Ben stopped Ed and laid it out for him, in his strong but honest farmer style: "Look, Ed – I'm sure you folks have some terrific technology that young folks are happy to use. But as for me and Nancy, we rely on our agent Jim for all of our needs."

Jim Doyle, Ben and Nancy's agent at Solar Corona, had built a long, trusting relationship with all of his clients, and Ben and Nancy were shining examples of Jim's success. He had grown up in Kansas on a farm and knew the business like the back of his hand. Jim and Ben had met many years ago, when Jim was just starting out as an agent. They developed a deep friendship with one another and Ben considered him almost part of the family. Many a Sunday you could find Jim having dinner with Ben and his family, catching up on all the news and talking business on the front porch after dessert.

After having a long conversation with Ben and Nancy, Linda and Ed got back in the car and began the drive back to the office. Ed was sullen for quite some time, but Linda didn't say a word. She quietly sat there and let Ed absorb what he just experienced. Finally, after what seemed like forever, Ed broke his silence. "Well, I'm not quite sure what to say after those two visits. In my retail experience, we never went out to visit any customers. But I bet if we had, we could have learned a lot, because I sure did today." Linda perked up and replied, "I'm curious, Ed; what is that you have learned?"

Ed replied: "I learned that I had a vision in my head that – if we put it into place – it would potentially have destroyed this company. I did not realize the power of the agent relationship and was totally fixated on eliminating them and letting customers use digital channels to serve themselves. What I learned today is that there is no one size fits all; we need to consider how to address the needs in a way that works for the different types of clients that we have."

Linda smiled broadly and said: "Yes, Ed. We need to look at things from our customers' perspective, not our own." Ed sat back in his seat and seemed to get lost in thought for the next hour while they drove back. Linda left him alone to ponder his thoughts.

Delivery Looks in the Mirror

"Linda, this was a great idea!," exclaimed Chuck as he walked into The Solar Corona executive suite at the United Center in Chicago. Linda had arranged to use the corporate suite to bring her team together to watch a Chicago Bulls basketball game that night. The suite was a private room with a great view of the basketball court, as well as being comfortably appointed with a full spread of refreshments available. "Yes, I agree," said Josanna. "We've all been working extremely hard and this is a nice reward!" Little did Josanna realize what Linda had in mind for their evening together.

"Steve, Howard – please come in," beckoned Linda. Both had been standing in the doorway to the suite, just a bit hesitant to enter, as if they had walked up to the wrong door and felt a bit lost. "You're in the right place," Linda continued. "Please – all of you – get comfortable; the game's about to start soon." Linda knew that all of her managers were basketball fans and watched games whenever they had the chance. She was counting on their love of the game to be the binding force to draw them all together to work as a team.

"Yes, you have all been working hard," said Linda echoing Josanna, "but I'm not sure all of our hard work is getting us to where we need to be." The mood in the room abruptly shifted, and Linda worried that maybe she started the discussion she wanted to have that evening a bit too harshly. "But – for the first half of the game – I want all of you to forget about work and have fun enjoying the game." Her statement seemed to ease the mood back to a safer place for the time being, but there was still some hesitancy in the air. Everyone settled down, though, and it was tip-off time. The crowd in the arena grew louder, and the room filled with anticipation for the game. Everyone settled into chairs in the front of the suite and began to lose themselves in the rivalry unfolding on the court.

At halftime, Linda asked everyone to join her at the table in the suite with some refreshments. "It's a close game," Linda started the conversation to break the ice. "But I think Chicago will win in the end," she continued. "Howard, what do you think?" Howard glanced around the table slowly, and then looked over at Linda: "Well, I think Chicago is going to pull it out, too – but I'm starting to think that's not the question you're really asking me, is it?" Linda looked directly at Howard and said, "Well, it could be there's more at stake here than just the game; wouldn't you agree, Chuck?" Linda slowly turned to look at Chuck as she finished her sentence.

Chuck was caught completely off-guard by Linda's question. He had been focused on the mini pulled pork sandwiches and French fries that he had filled his plate with before sitting down at the table. Startled, Chuck looked up at Linda and said, "Uh – sorry Linda, what was that again?" But before Linda could reply, Josanna stepped in and said: "I think what Linda is getting at is the fact that she's worried about the outcome of *our game*, isn't that right, Linda?" Linda looked over at Josanna and smiled. Now, Steve—feeling a bit removed from the point being made—pulled his chair in a bit closer and asked: "Our game? – I'm not 100% sure I'm following where this is going."

"Why do you think Chicago is going to win the game?," Linda said, seemingly asking everyone at the table together. Howard, Chuck, Steve, and Josanna all looked at each other. Chuck, wanting to redeem himself a bit, was the first to reply: "I think their three-point shooting has been awesome in the first half, and if they can continue, that's what's going to win the game for them." Steve almost immediately shot back at Chuck, saying "Chuck – it's defense that wins games. If they don't stop Miami driving to the hoop, they'll never be able to pull it out."

"Teamwork," Howard said softly, almost as if in a trance, looking down at the table. Steve looked at Howard and said, "Well, of course teamwork is important – that goes without saying." "Well, it's interesting that you consider it that way for the basketball game, but not back at the office," Josanna retorted. Steve glared at Josanna, but before he could say anything, Howard jumped in to her defense and said, "Josanna's right, but it's each one of us that needs to rethink what should be obvious." Howard continued: "And I'm probably the one who has been most guilty out of the group. I had my suspicions as to why Linda brought us all here tonight. She's using this game to send us a message. We need to play better as a team."

Chuck, always just a half-step behind, looked around the table and said, "But we're not a team – we're a bunch of managers." Josanna looked at Chuck and smiled at his remark, since she knew it was coming from a position of truly not knowing where this was all going; but then she turned to Steve as her expression went simultaneously from smile to a scowl as she said, "That's the point! We've been so consumed with 'managing' our own vertical silos that we've lost sight of the bigger picture. And the bigger picture is winning the game – not for ourselves as individuals – but for our customers as the provider they've come to rely on."

Linda finally began to speak. "If anyone is to shoulder the brunt of the blame, it should be me," she said firmly. "I've been the one that has created the structure that we have each come to be part of, and when you put great people in a poor design, it's hard for them to succeed."

Linda and her team continued to discuss the situation in more detail until the halftime break came to a close. As the second half of the game was about to begin, Linda looked at her team and said, "We've cracked open a pretty big egg, and I'm really encouraged how each of you has been willing to discuss your role in this. While we watch the second half, let's do two things: first, let's see what we can learn from both teams as they each try to win the game, and second — let's all hope Chicago wins!" Linda asked everyone to help themselves to more refreshments, and to get comfortable for the start of the second half. She knew there was much more at stake than just the ballgame going forward.

Doing the Analysis

Acknowledging the Past

Linda had one week to go before the CEO's deadline of coming back with a solution to his questions. She knew she needed to have all of the executives on board before getting back to Davis. If there was one thing that Linda had learned through her years at Solar Corona, it was a team proposal that was required when presenting something so dramatically different from the way the organization had been running for the last several decades. But Linda knew that she needed to pay respect to the past as she proposed a new future.

Linda proposed a two-day workshop to all of the executives and managers that she had met with in the last two weeks. She wanted the first day to be one of both reflection and stimulus, while the second day would be reserved for synthesizing everything discussed on day one, as well as constructing a proposal to answer Davis' challenge delivered during the budget meeting. Because Linda had taken the approach that she did to get all of the key players primed for the conversation, she was cautiously optimistic that they might actually be able to achieve the goal that Linda was trying to meet.

Linda's invitation went out to everyone she had met with, and to her surprise, everyone accepted within a matter of hours on the same day. She had arranged the session for two full days on Wednesday and Thursday, knowing that people would be tired by the time a Friday afternoon rolled around.

Wednesday morning came quickly. Linda went to the training center early that morning to make sure everything was set for the session. When she arrived, she found Will Jenkins, the company historian, already there in the room. Linda had asked Will to present material covering Solar Corona's past, not only to acknowledge and thank everyone in the building for creating such a great company over the decades, but also to remind everyone in the room of what they were in business to do—which was to help their customers.

"Will, I really want to thank you for participating in this session today – I'm really looking forward to everyone hearing 'our story'," Linda said, reaching out to shake Will's hand. Will, his hand meeting Linda's, replied, "Linda – it is my pleasure. I'm excited to be a part of what you have planned for the group." Linda had explained the whole story to Will a few days ago, so that he would understand what she was trying to accomplish. She asked Will to provide everyone with a historical view of how Solar Corona had always been there for their customers, even as time had gone by and the challenges they faced had evolved.

Will had pulled together old photographs, film snippets, and audio recordings and created a video montage of the history Linda had requested. He had constructed segments of material, each focused on the challenges, successes, and evolution within each decade from the beginning of the company in the 1900s to the present day.

There were some decades that reflected hardship and pain, especially in the early days of the 1920s and 30s. Other segments of the montage reflected the enormous promise of growth and prosperity in the 1950s, after the war. But regardless of the specific events covered in the segment, what was clear was the emphasis on Solar Corona helping its customers through tough times and seeing them through to help them prosper. What began to resonate with everyone was the pattern of Solar Corona growing in response to helping customers through the challenges that they experienced in life. But what also became clear is how that response and related growth had slowed down in the later decades.

What permeated the later segments of the montage were stories of the acquisitions that Solar Corona had made throughout the 1980s and 90s and how that had changed the company, including the demutualization event. The change in demeanor in the room was almost palpable, as the audience went from feeling proud and exuberant about the early stories, to feeling downtrodden and drained as they watched the more recent parts of the story. But this was all part of Linda's plan. She knew she had to reach not only their heads, but their hearts as well, if she was going to sell them all on what was to come.

The video montage ended, and Linda thanked Will for all his hard work in putting the materials together. Linda went to the podium in the front of the room and addressed the crowd. "I hope all of you found something meaningful in that montage," Linda said to the group. "I think it speaks to who we once were and what we have now become." The room, which had been in a bit of a buzz, fell silent. Linda held her pause for what seemed like ages, letting everyone dwell on what she had said, and what it meant to each of them.

Linda—without saying it directly—had just thrown down the gauntlet. After a few moments, Linda spoke again: "I think each of you should reflect on what you've just seen, take a break and come back in 20 minutes ready to talk about who we want to be."

Exploring the Future Together

What Are Our SWOTs?

After 20 minutes had passed, Linda went back up to the podium and waited for the group to settle into their seats. She started up her laptop and connected it to the overhead projector. On the screen were four letters in very large text. They read "S W O T," split out where each letter was in the top corner of one of each of the four quadrants that the page was divided into.

Linda waited till the room went silent, with nothing but the hum of the projector in the background.

"Can anyone tell me what these four letters stand for?," asked Linda as she pointed to the screen image up on the wall. Alice Gray, who had attended one of Linda's earlier sessions and had also arranged the field trip for Linda and Ed, raised her hand. "Yes, Alice – do you know what these letters stand for?," Linda asked. Alice said "Yes, Linda, I do. It stands for Strengths, Weaknesses, Opportunities, and Threats. It's a technique that I've used in the past to help us identify both the internal – Strengths and Weaknesses, and the external – Opportunities and Threats – that we need to examine when analyzing a problem space and to help us understand how to respond." "Excellent, Alice! I'm going to ask you to lead out first breakout group this afternoon on examining Solar Corona's SWOTs when it comes to becoming who we want to be as we go forward. But I'll explain that more in a moment," said Linda as she then switched images on the display screen. What came up next was a bit of a surprise to everyone in the room.

What Are Our Value Streams?

It was an image of a river of flowing water that kept getting interrupted along the way by great big boulders. Linda again asked if anyone in the room knew what this picture represented. This time it was Josanna Shipton who raised their hand. "Linda, I think what you're trying to introduce is a 'value stream,' is that right?" Linda said, "Yes – that's correct, Josanna. But do you know what the boulders in the river are meant to represent?" Ann Demming, the Chief Marketing Officer, shouted out, "I know – those are meant to be interruptions in the flow of value; just like they are stopping the water from reaching the end of the river, there are obstacles that exist that slow down or even prevent value from reaching our customers."

"Precisely, Ann," replied Linda. Linda continued: "You see, you can think of each one of the boulders as the various departments and systems that are required as a customer moves from identifying their initial needs for insurance to when they actually need us to help them through a claim process when something unfortunate occurs. The challenge for us is to keep the boulders as small as we can so as to not interrupt the flow of value to our customer."

It's at this point that Ed Boderman, head of channels, joined in and added, "Maybe the key to that is to match the size of the boulders with the size of the flow – the smaller the flow, the smaller the boulders – the larger the flow, the bigger they can be without causing too much disruption." Linda was smiling now, seeing the room start to rebound from the discomfort earlier in the morning. "That's great – I think we are all getting the picture – no pun intended," said Linda, as she glanced back at the image on the screen. "If you are willing, I would like to ask Ed, Josanna, and Ann to colead the second breakout session in the afternoon on value streams. Ed, Ann, and Josanna all nodded their heads in agreement.

What Forces Are at Play?

"OK, then, we're almost at our lunch break," said Linda. "But before we break, there's one more important tool that I'd like you to be armed with when you go into your breakouts this afternoon." Linda went back to her laptop and switched slides to show her third and last image for the morning. It was a picture of a big rectangle in the middle of the slide with the words "Solar Corona" inside the rectangle. To the left were arrows "pushing" the rectangle forward, while to the right of the rectangle were arrows pushing the rectangle "backward."

"Can anyone tell me what this image represents?," asked Linda. Oscar Dayton, the CFO, shouted out: "Yeah – that's our margins getting squeezed!" The room burst into laughter at Oscar's pronouncement, but then as the laughter subsided, Howard Langford in IT said, "Although I can appreciate Oscar's sentiment, I believe it's meant to be an image that is depicting a force field analysis. When we started the emerging markets group in IT, we used force field analysis to help us identify some of the areas that were holding us back in addressing the needs of those newer potential customers of Solar Corona. The arrows to the left represent those forces that are pushing us forward, while the ones on the right are those holding us back."

Linda replied: "That's right, Howard. And now we need to do the same thing, but not just for emerging markets; can I please ask you to lead that breakout group this afternoon?" Howard nodded in agreement and Linda knew that now was the time to bring it all together and then send the teams off to run their breakouts.

"What we've discussed now since the video montage have been three tools, each designed to help us understand how to create the next version of Solar Corona. The SWOT analysis will help us understand the internal and external playing fields that we are operating in and provide us a way to think about how to both leverage our strengths and opportunities and minimize our threats and weaknesses. The value streams will provide us a way to turn our individual, vertical 'boulders' into a flattened, horizontal stream that better enables the flow of value to our customers. And last but not least, our force field analysis will help us identify those winds that are blowing us toward our new future, in addition to the headwinds that are trying to keep us where we are today," Linda explained, and then she provided the instructions to the group and sent them on their way for the remainder of the afternoon.

Creating the Coalition

Cocreating the Future

When the group came back together again for day 2, the excitement was palpable. Everyone was invigorated by the breakout sessions held during the afternoon of day 1. Linda went up to the podium in the front of the room and welcomed everyone back for day 2.

"Good morning, all," Linda said enthusiastically, looking out over the group. "I'm excited to explore what you all came up with in your breakouts yesterday afternoon. But before we get started, I would just like to lay out a few ground rules for our discussions this morning."

Linda continued, "I'm sure as you all went through the sessions yesterday, you were able to see that each and every one of you – regardless of position in the company – are able to contribute in a meaningful way. Some of you have had more experience in certain areas, not by virtue of your title, but by what you have been fortunate enough to be exposed to and work on throughout your career. This is why it is important, as we move forward today, that we leave any titles at the door. We will be discussing what a future operating model needs to look like here at Solar Corona, to be proposed to Davis McMillan, our CEO, next week."

The remainder of the day was spent discussing what the future operating model needed to look like for Solar Corona. Linda instructed the group to use all of the materials from day 1 and think outside the box, leaving their current understanding of the business behind for the day.

"What's important in our work today is to be genuinely selfless as we think about the kind of future operating model that we want to have here at Solar Corona," said Linda. "And we can't get hung up on every little detail right now; we need to take broad strokes with our brushes and get at the heart of how the operating model needs to transform. My advice is to simplify things into key phrases that will resonate with Davis, and eventually the rest of the organization."

Linda was fearful—afraid that they just wouldn't come through with the right level and content of a proposal that would get Davis to react positively. But, she knew she needed to let the group own it—because in reality, that's who really did.

Proposing a New Operating Model for Solar Corona

The meeting with Davis was set to start promptly at 10:00 AM that morning. Linda was nervous, but excited at the same time. She had managed to get to the right mix of people and have them generate a proposal that they would actually present to Davis. The approach was much more than what Linda had been asked to do, and she was worried as to how Davis would react.

As the participants settled down, Linda got up out of her chair and addressed the audience: "First of all, I just want to thank all of you who have participated in all of the various sessions we have held over the last month. Davis, the work you will hear today is the work of us as a team; everyone has come together to create a proposal that we all believe to be an approach that puts the company on a new journey – a journey that will eventually lead to increased value for our customers, in turn which will help Solar Corona to prosper." Linda continued, "I've asked various executives to present specific components of our proposal, so that you can hear from them and see that it is what they have created and now believe in, as well as probe them with questions so you get a feel for the depth of their conviction. They are going to present a set of transformations that must occur in our organization – new ways of looking at certain aspects of our business that enable Solar Corona to evolve and grow. So, without any further delay, I'd like to ask Ed, Ann, and Josanna to start us off."

Ed, Ann, and Josanna stood up and went to the front of the room to present their story. The key element of their story—using the discussions and materials from the two-day workshop on value streams—was on seeing the customer as the focal point for everything that Solar Corona does. The three presented discussion materials that had been collected to tell stories of too many processes where information was simply "thrown over the wall" to the next department in a series of handoffs that went on forever. This silo mindset had been built up over years and years, with many of the acquisitions adding to the long chain of handoffs. Solar Corona had lost sight of the customer waiting at the end of the line.

"We need to re-evaluate the different types of customers we have, the products and services that they are looking for, and how best to deliver those," said Ann. Ed continued the thought: "And we need to deliver those via their channel of choice – whether that's digitally or through an agent." Josanna completed the sentiment by saying, "We must put our customers first – in any of the thinking we do. We need to be sensitive to their preferences, understand what they want and how they want it, even before they realize they need it!"

Davis was taken aback by Ed's admission that it was the customer who should decide the channels that were used to deliver the products and services that they needed, not Solar Corona employees. This was contrary to everything that Ed had been espousing since he had arrived at the company. Davis was intrigued; if Ed was capable of such a turnaround, Davis thought to himself, maybe the rest of what's coming is worth listening to!

Alice and Chuck were next to present. Their topic built on the first talk delivered by Ann, Ed, and Josanna on value streams, and focused on the "boulders" placed throughout the stream. "We have a huge shift in thinking that we need to make," Chuck began. "We need to move from being in competition to being in cooperation with one another." Alice started to add on to Chuck's words: "In the previous discussion led by Ann, Ed, and Josanna, we talked about how the customer has to be everyone's focus. When we change that focus, it means the various silos that we've created have to move from being in competition with one another – competition for funding, for people, for technology, et cetera – to being in cooperation with one another, all aligned to deliver the best products and services to the end customer – the one that's paying for it. We can no longer afford to let each silo focus on trying to optimize itself; we need to optimize the whole value stream."

Davis almost fainted. He was utterly amazed to hear Chuck—who had fought the creation of the emerging markets unit and bringing in Howard to run it—talk in such a selfless manner! Now he knew that Linda had really accomplished something over the last month, and he was beginning to believe in it himself. But he still hadn't heard an answer to the core question of the budget and how they were going to use what they just presented as a way of saving money.

Davis did not have to wait long. The next set of material was about to be presented by Oscar and Steve. Oscar got up to the front of the room and started to speak: "Well, I guess I am probably the most surprised out of everyone here," Oscar began. "Davis – I know this all started with me coming in here a month ago sounding the alarm, and – to tell you the truth – I'm glad I did! The teamwork and innovative thinking that I have seen over these last four weeks have been amazing! I've learned so much, and from so many talented people, that I have a whole new perspective on our budget. We have been looking at it through a certain lens that treats everything as a cost and we've been conditioned to react to those costs by trying to control them. What I've learned from my colleagues is that it's not about the cost per se, it's more about the value." Steve jumped into the discussion: "That's right, Oscar. And just like you, I've been running our infrastructure with a sole focus around cost, without thinking about the value it needed to provide. We've been entirely focused on taking cost out, but we've forgotten about the value we need to provide to our customers. So much of our business – especially through our acquisitions and expansions – is rooted in providing shared services across our various line of business. But what we've lost in that complex view is that not all of those lines of business have the same return when it comes to the revenue that they bring in. That's the 'value' part of the conversation. We need to rethink our investment strategy so that the value streams that generate the most margin are afforded a way to invest and grow without getting bogged down by an across-the-board budget cut."

Linda now took over to wrap up the presentation and to ask for an approval from Davis. Linda started by recapping what each of the various sections had covered and offered a perspective that summarized the main theme of the day. "Ultimately, it's about Solar Corona making a switch from utilization to flow," Linda said. "We need to stop worrying about the utilization of all of our people, machines, systems, and so forth, and get focused on the flow of value out to our customers. It's not that cost isn't important; it's just that it needs to be looked at with a different lens: a lens that places more importance on overall value as opposed to simply managing cost."

"Davis," Linda said, looking directly at him now, "We need a new way to operate: a way in which we quickly test new products and services in our different value streams to see if they work without spending a lot of time and money up front; a way to work across boundaries to pull together various capabilities that don't belong to one particular silo or another, but belong to Solar Corona as a whole; a way to sense and respond to what's happening in the market so that we invest in the things people want and divest in those that no longer provide them any value."

Linda continued, "Now, I know this is a big change from the way we operate today – and it's not going to happen overnight. Quite frankly, I'm not sure if you even believe all of us when we say it will work; you certainly have the right to be skeptical. That's why our proposal is not to change the world overnight. Our proposal is to start with a meaningful pilot effort. Something that is small enough to manage, but large enough to make a difference and understand what a larger-scale shift would mean."

Davis looked at Linda, and then slowly looked at everyone around the room. He was proud of the team, but at the same time knew that what they were suggesting meant a long, tough road ahead for Solar Corona. But Davis also knew that, in order for the company to thrive in the years ahead, it needed a radical solution that was out in front of change as opposed to constantly trying to play catch-up to it.

"OK everyone," said Davis as he rose from his chair. "Let's try a pilot."

Learn from Linda—Look in the Mirror

"It happened again," Linda groaned to her colleague. "Things seemed to be starting off just fine in my first meeting with the business executives since there was a common awareness about the company problem. But then I was frozen out from further communications." Over and over Linda, heard the executives say the same thing after she met the executives: "I understand that our company has challenges, and you are passionate about agile ways of working that delivery has been using, but I don't see how this has anything to do with us." The executives had cut their meetings short and Linda had not been able to get another meeting.

What Linda experienced was the negative consequence of her initial conversations having too much focus on accomplishing her goals for change, instead of the executives' business goals. In her case, the executives deflected what they perceived as self-serving advances to her own agenda.

Linda's colleague gave this advice: "Take the time to discuss how the executives do things under their current operating model and lead the conversation so they can draw their own conclusions and have the knowledge that they are part of the company problem."

Linda went back to these executives and offered to have a discussion instead of a presentation. She proceeded to her next meetings with a collaborative tone and began converting executive intrigue into trust by discussing their operating model and their perspectives in dealing with the looming company problem. Rather than blurting out what she wanted to share, she allowed them to explain their operating model and then led them in identifying the root problems that could be addressed. Linda once reflected, "During these conversations, I got the executives to look in the mirror, put egos aside, and acknowledge that their existing operating model will not solve the company problem."

CHAPTER

8

Shining the Light on Solar Corona

Finding the First Problem

Sorting Through the Mess

Linda was happy that the meeting had gone so well. Now came the tough task of trying to identify a pilot where they could demonstrate all the transformational change that had been presented to Davis in a way that would actually provide value to customers. Linda asked Chuck, Howard, Josanna, and Steve to bring together all the information about the current portfolio of work that they had in IT, so they could begin to sort through any of the opportunities that might be suitable for a pilot.

Linda opened the meeting by saying, "I'm open to suggestions, everyone; I don't have a particular pilot in mind and I'm not sure what the best way is to identify the right pilot." Chuck responded, "Well, based on what we discussed in our workshop, it seems to me that we need to find the right value stream first before we do anything else."

J. Orvos, *Achieving Business Agility*, https://doi.org/10.1007/978-1-4842-3855-4_8

Howard looked perplexed, and started speaking to the group: "Chuck, I agree with you but I'm not sure how we even have enough capacity to undertake a pilot when we have all the other commitments lined up for the rest of the year." Steve echoed Howard's sentiment: "I agree. I have a series of infrastructure projects defined to support all the other initiatives that are trying to get done this year. How are we supposed to interrupt all our plans to squeeze in a pilot?

Linda felt the anxiety that was building in the room. "OK, everyone, I get it. I understand your frustration and fear," said Linda. "We have several commitments already defined, and I'm sure no one wants to disappoint any of our customers and stakeholders that are waiting on the results of those efforts. I think what's important in identifying the pilot is understanding the commitments that we have and at the same time identifying where the value of using our new operating model could help us deliver a better outcome, with a better approach, in such a way as to keep the commitments that we have made."

Josanna spoke up: "Ahhh … I think I know what you mean, Linda. When we did our presentation to Davis we talked about ways of putting out small pieces of functionality to see if they're what the customer wanted before building an entire solution. Maybe we can find an area where we have an investment that's planned for the right value stream and we change our approach to use a new delivery model to test aspects of the solution before we spend too much money and time on delivering the entire product." Chuck jumped back in and said, "OK – I can buy that. But how will we know what to build as a test case before building the entire product?"

Linda said to Chuck, "Do you think we know enough about what our customers want to provide value to them incrementally, while at the same time increasing our margins in order to solve our financial issues?" Linda already knew the answer was no, but she needed the group to get centered again on the new operating model that they all proposed to Davis. Chuck replied, "No – I used to think I knew what our customers wanted, but as times have changed, I feel like I'm not plugged in enough to their needs."

Linda replied, "Chuck don't feel bad – I think we're all in the same position. Over the years, this business has gotten much more complex, and we need to pull in many diverse perspectives to develop products and services that provide value to our customers. And, quite frankly, it must start with the customers themselves."

"So, one of those perspectives needs to be marketing," said Josanna. "Yes," Linda replied, "but not in the traditional sense of how we think of marketing today. The way we market things today is by using a 'sell-it-and-they-will-come' approach. We need to truly put the customer first by asking them what they need, not by guessing and then hoping we got it right."

Steve spoke up: "I see where this is going, and I'm really nervous about it." "Why is that, Steve?," Linda asked. "Well, if we're going to start with asking customers what they want, and then start to build it a little at a time, how will my team know what kind of infrastructure will be required to support the effort going forward? Today, we have a lot of forms and processes that provide the data we need to understand the requirements, so we can build out the supporting infrastructure, which takes time," Steve declared.

"We have to stop thinking in silos, Steve," Linda replied. "All those forms and processes and then the wait time to get all the equipment in – they're all big boulders in the value stream. We need to figure out a way to deliver the infrastructure incrementally as well, in a just-in-time fashion. We can't afford to wait till we know everything, because we never will; but at the same time, we can't build something that costs a fortune, just to find out that our customers don't want to use it."

Steve reluctantly acknowledged that Linda was right. Everyone in the room was beginning to understand in more detail now exactly what the transformations would mean for each of them. It was going to be painful, but they all understood and agreed that the pain would be worth it if it meant that there wouldn't be any arbitrary budget cuts. Everyone agreed it was much better to control their own destiny.

The team discussed things for a while longer, and agreed at the end of the session that they needed to bring the rest of the team in to talk about how to select a pilot. Having gone through some additional learning and "a-ha" moments themselves, they knew that the rest of the participants would need to personally experience these as well.

Gaining Consensus on Where to Focus

It was now two weeks after Linda had brought her IT management team together, and all of the other participants had met to experience the same sort of epiphany that Linda's team had undergone. Linda had arranged for the larger group to come together now to discuss how to move forward with the pilot.

Linda opened the session: "Well, everyone, we've all – I think – realized that we need each other to figure out what the right pilot is; there's no one function, person, or team that can do this by themselves." Everyone in the room nodded in agreement.

Ann Deming, the Chief Marketing Officer, offered up some input she had brought to the meeting. "We've been having meetings with some of our clients," Ann began. "I think, based on their feedback, that I may have identified an area that is a potential fit for the pilot. I've also sat down with Oscar and gone through some very preliminary numbers, and I think it will fill a gap that exists in our offerings, and not only make us money on the offering itself, but maybe drive persistency with our existing customers as well as growth with new customers," Ann continued.

Ed spoke up: "That's sounds promising, Ann. Tell us more about what you've come up with." Ann replied, "Well, it's very preliminary, and there's a lot of work to do to figure this all out, but here's what I've identified so far. We held some focus groups with customers – including the ones that you visited, Ed – to discuss what their needs were. What we got back was a very strong indication for us to provide transportation insurance for goods that are being transported from customer farms to both customer and noncustomer small businesses. For example, Ed, when you visited the Bakers – both Ben and Nancy and Mark and Gail – you heard about the small-batch distillery that took the grain from Ben's farm to process into bourbon, right?," Ann asked.

"Yes – I remember," said Ed. "Well," continued Ann, "it turns out that we insure the grain while it's on Ben's farm, but once it goes into the trucks for transportation, it's at risk. Turns out that this is the case for a lot of our customers, and they are looking for inexpensive ways to insure their crops during transportation."

"But aren't there transportation companies that provide that type of insurance?," Ed asked. This time, Oscar spoke up: "Yes, there are, Ed. But their prices have been steadily increasing and the types of events that they cover getting more and more narrow. Our customers have found themselves in a quandary, especially when they are in the situation like the Bakers. They don't want to have to increase the price of the grain to cover the cost of the insurance, because – since it's family they are selling to – they are already operating on slim margins. They are looking for ways to decrease the cost even below what they are paying today."

"So," Oscar continued, "if we could come up with a profitable offering that beat out the transportation insurers, not only would it make us money directly, but it would strengthen the core foundation of helping our existing customers and help support a more multigenerational outlook for persistency."

"That sounds like it would deliver significant value, but how do we build something like that incrementally?," asked Chuck. "Well, when we build analytics, we always start with the basic functionality and then add on the specialty types of analytical processing that our users are looking for," said Josanna. "We know from experience that they won't know what they want until they start playing with the data and using it to actually ask questions and learn from their studies."

Then, Ed asked, "So, what you're saying is we need to figure out what customers want as the sort of 'base' set of functionality and then – after they start using it – we get more feedback from them on what they want next and then build that?" Linda saw an opportunity here to help establish some common language for the group. "That 'base set' of functionality has a name that is used consistently in the industry," Linda offered. "It's called the minimum viable product – or MVP for short. You see, it's a bit more complex than just carving out a basic set of functionalities," Linda continued. "It's truly identifying the bare minimum that the customer will need to be willing to use the product or service, no more and no less."

"And do we need to be able to make a profit on that MVP?," asked Oscar. Linda answered with an emphatic no. "The purpose of the MVP is to determine if what we've come up with actually will help solve the customer's need and that it actually technically can be done," Linda explained. "So long as we think that – at some point in the development of the offering – it will begin to be profitable, that's enough to drive us forward to learn more. Remember, the only one that can tell us if we got it right is the customer. We need to work together with them on a journey we take on building the offering together."

"So, how do we work with the customers – do we give them a list of things and have them tell us what they want; or do we come up with something and see if they can use it?," asked Josanna. "Well, if we look at what it takes from the point of view of the client, that's a great place to start," replied Linda. "And, remember, we're looking for a minimum viable product, not something with all the bells and whistles on it. I think we should look to Ann to help us work with clients to find out what they absolutely need, as well as work with Ed, Chuck, and all the others to find out what can be built or that exists already that can be leveraged. What we're looking for, in other words, is the set of capabilities that will be necessary to minimally support the need that a customer has to insure goods for transport. We need to start by focusing in on that simple offering of covering the goods being shipped so they are protected."

"That's right, Linda," said Ann. "I've been studying the market for years and thinking of ways to differentiate ourselves. I'm not saying I know what the client wants: what I am saying is I know the various ways to categorize how we investigate opportunities to make the offering better, faster, and cheaper to use, all while improving the customer's overall satisfaction."

"Well, I hope that investigation also identified ways to figure out not only what to deliver, but how long it will take," said Howard. Ann looked a bit confused and said, "What do you mean, Howard?" "I'm talking about knowing what the customer wants is only half the problem," Howard replied. "The other half is figuring out how you're going to deliver it in a short enough time frame to keep them happy so that they don't get up and leave."

It's at that moment that everyone in the room had another epiphany. They realized that the customer half of the equation was driving what the need was, but then there was a Solar Corona half of the equation that encompassed delivering that need. And this needed to be a repeating cycle, not just a one-time effort. Simply put, Solar Corona had to manage finding the most valuable thing to deliver in the shortest amount of time for its customers, and then do it again and again and again—until there was no more value to one or both parties.

Carving Out the First Slice

Linda felt good about the progress the group had made in the session that day. There were a few more topics to cover, but Linda felt close to having enough of an initial consensus on what to choose as the pilot. what was left to discuss was precisely what everyone would consider to be an MVP that they could roll out and then build upon as they received feedback from the marketplace.

There was an air of caution in the room as they discussed what would be offered initially. "I think we should push the self-service model as a starting point," declared Ed. "It just seems easier than involving the agents. And it puts the responsibility for entering in all of the information on the customer, so it's less involvement – and expense – for us long-term."

Linda could see Ed slipping back into his old ways again, but she knew that this would ultimately all get worked out by getting the feedback from the customers. She let the group continue.

"We can limit the MVP to existing customers – that will help us not have to worry about building any new customer profile capabilities – at least not as part of the MVP," offered Ann. Oscar chimed in, "We can also limit it to certain farmers and ranchers that we have good relationships with – you know, it's always good to start with friends." Ann laughed out loud and said, "Oscar – we're going to make a marketer out you yet!"

The group continued to discuss various other ways of expanding and contracting what would eventually drive the scope of the MVP, and after another 90 minutes, had reached a degree of consensus that they felt they could now start with.

Taking Action—Establishing the Pilot at Solar Corona

Organizing Horizontally

Using the high-level MVP scope that the group had agreed upon, they now had a good understanding of the various departments and functions that would need to be represented on the pilot team. Everyone agreed on the importance of having the full value stream represented, and to staff the team with members that would be dedicated to the pilot, at least through delivery of the MVP. Based on the success of the MVP, then the structure of the pilot would be evaluated as to how to either move forward.

There were team members from a multitude of different departments in the end-to-end value stream. There were people from marketing, finance, sales, channels, delivery, claims, the service center, and many more. Anyone that was viewed as having any involvement in defining, creating, selling, and servicing the offering was included in the team. This was a very new concept for the various parties involved, as Solar Corona had operated in silos for so long. It was very hard for people to understand how this huge team was going to operate without the perceived structure and control provided by their functional silo.

Not only were the functional silos left at the door when a participant joined the team, but so were their title and specific role within the organization. Of course, the various areas of expertise were acknowledged, but the rank that was typically associated with these was ignored while on the team. Before the team even started to work together, this concept proved extremely difficult for some people that had become accustomed to managing by rank and position within the organization. Some very painful discussions were had as participants were identified and interviewed to be part of the team. Certain individuals that seemed not to be able to wrap their heads around this way of working were left in their existing roles and not invited to join the team.

It was extremely important to focus on these aspects of the pilot team structure and composition, since so much success was riding on the ability to remove the "boulders" from the stream. If Solar Corona was going to be able to work quickly and efficiently, it needed to throw off the trappings of the past, which meant no existing organizational silos and associated "walls" for things to get thrown over for the next department to worry about.

Customer Centricity in Action

Susan and George, the scrum master and the product owner that Linda had called on to help in prior conversations, were asked again to play those same roles on the pilot. Both were fairly knowledgeable in their respective areas, and also had the right mindset of facilitating the entire team to work toward the outcome together.

George knew that the very first thing that had to be done was to establish the relationship with the customer. This was critical to drive what the offering would need to look like and how best to decide on value as they incrementally developed it. George was looking forward to the first day of meetings with the Bakers, whom Davis had reached out to personally to ask them if they would participate in this great experiment.

Ben and Nancy as well as Mark and Gail showed up bright and early at Solar Corona's office on the day that George had invited them to come in on.

George greeted them at the door and led them to the team room that had been created for the pilot, where everyone was waiting to meet with them.

"Wow!," exclaimed Ben, seeing all of the people in the room: "I didn't realize we were this important!" Everyone chuckled a bit, and then George said, "Well, Ben, you and Nancy and Mark and Gail represent the future of Solar Corona. Your ability to be successful helps us be successful."

After some continued introductions, the group got down to business. George presented the product offering ideas and MVP that the group had developed as a starting point to the Bakers. Some members of the team had been a bit reluctant to bring in the Bakers so early on in the process. They felt that SCI should have started to build the offering first and then bring in the Bakers to present it to them and gather their feedback. But George, with Susan's help, coached the various team members on why it was so important to start the product vision with the customer view.

After thinking about the MVP presentation, the Bakers slowly began to respond. "You know, I like the idea of getting the transportation insurance from the same company that protects my crops from the get-go," declared Ben. "But – as you guys well know – I like dealing with my agent Jim Doyle. I'm not a fan of all of this technology stuff, and I don't want to have to get on a computer to do all the details about the shipment. Can't Jim do that for me?"

This was a scenario that they had not considered for the initial MVP. It meant that they would have to provide some sort of functionality for the agents to enter in the shipment details in order to get the insurance rates and establish coverage for the shipment. But George knew that if Ben brought it up as a need, then there would be other customers that would see it that way as well. He knew that SCI had built its reputation throughout the years on the good service provided by agents, and that there were still plenty of customers that got regular visits from them.

"I think that's a reasonable request," George replied to Ben. And over the next two hours, the Bakers talked with the team about a diverse set of needs: from pricing to access, coverage, and claims, they covered it all. When everyone felt they had covered all they could until they actually started to provide the offering and the Bakers started using it, they summarized all of the items on the new "wish list," which George referred to as their "backlog." They had actually generated a list of 47 items, which shocked everyone on the team. George thanked the Bakers for coming and participating in the session and that they would be hearing from SCI when they had something that could actually be offered to insure the deliveries.

After the Bakers left, the room was abuzz. First, they were surprised that most of the items in the backlog were things that had not been considered in the MVP discussions, regardless of when they would be delivered. Second, the team was amazed at some of the innovative ideas that came up during the session. For example, Mark Baker, concerned about the price of the grain (he knew that even though it was his own father selling it to them, Dad would pass along any extra cost in the price!), came up with some fantastic ideas of how to lower the cost of the coverage. He figured that if there was a way that SCI could "track" the delivery vehicles so that they only used certain roads that were known to be less traveled and therefore safer, that it might provide them an opportunity to save on cost since the risk was lower.

George was really pleased to see how the discussion had invigorated the room and imparted a sense of connectedness and urgency to all of the team to help solve the Bakers' needs. But he also knew that they had opened a door and walked through it, and there would be no turning back from this point forward. They had now established—albeit still a bit fuzzily—an expectation with the Bakers that they would be working with SCI on a journey together.

Defining Success

After the room calmed down from their newfound excitement, George got up and began to speak: "For most of you, that's the first time you've probably dealt directly with a customer in any capacity. It's an amazing feeling to be connected directly, to hear them express their needs and get to interact in a way that brings you newfound knowledge." George continued: "But now the hard part begins. Now we've set an expectation and we need to deliver. We need to be able to measure our success in delivering on that expectation. Remember, we've agreed that we are not going to build everything at once – so we're already going to disappoint the Bakers and other customers that need this offering. We have to find a way to measure success *along the way*, knowing that it's going to be a journey, not a destination. And that journey is something that we are going on together with our customers, so we need to be able to measure their success as well as ours."

"We need to change what we measure, don't we?!," exclaimed both Ann and Howard almost in unison. But Oscar asked, "What do you mean? Don't we still need to measure how many customers are signing up for the offering, how much revenue it's bringing in, how much it's costing to provide, and so on?" George replied, "Oscar – you're right, we do. But think about where we are in the process. We don't have a full-fledged product offering ready to sell yet. We need measures that will help us steer the development of that offering down the right road. That's part of the power of not building everything all at once. You get to frequently deliver, measure, course-correct, and continue."

"We need to define 'leading indicators'," added Susan. Oscar looked at her and asked, "What's a leading indicator?" Susan explained: "Since we're going on a journey, we need a way to measure our progress in such a way that it tells us that we are heading in the right direction *before* we reach our destination. If we wait to look at sales numbers and NPS scores, it's going to be too late to do anything about them. That's why those are called lagging indicators. They provide information after the fact but aren't very helpful when you're trying to steer your way to success."

Oscar said, "Oh – I see; it's sort of like my gas gauge on my car, coupled with the little electronic gizmo that tells me how many miles per gallon I'm getting. Together, they are leading indicators of when I'm going to run out of gas." "That's right," replied Susan: "Knowing how fast you're burning gas will help prevent you from running out and getting stranded. We need a similar set of measures that will help us understand how well we are doing, both for the customer, as well as for the company, before we 'run out of gas'."

"So, what ideas does anyone have as to the right set of leading indicators that would tell us how we are doing along the journey – and that we're headed in the right direction?," asked Oscar. "I think we have to look at the journey and derive indicators that are meaningful at certain points along the way," said Ann. "One example would be testing the MVP, especially in light of all the feedback we got from the Bakers. Maybe we put up some material on our website to begin to gauge the interest for the offering – kind of what we did with the Bakers in the room, only do it online for a larger audience." "That's a great idea," exclaimed Howard. "We can get all sorts of information that we can use to help steer things like channels, pricing, and customer demographics."

"And then, as we roll out the offering, we can measure how much certain aspects are utilized," added Ann. "For example, we heard loud and clear that Ben wants to be able to have his agent Jim Doyle involved to help him, but we also know we need to provide a self-service capability to those customers that have a preference for arranging for coverage on their own without getting their agents involved. We can measure how much each is being utilized before we invest more heavily in either; this way what we invest in is driven by actual customer demand."

Linda looked at Oscar and smiled. Oscar knew why Linda was smiling at him and he smiled right back. The two of them were silently recounting the initial conversation that Linda had with Oscar around investing in the things that had the greatest return. Oscar was now beginning to see it in action and was pleased as punch! "So, Ann," Oscar said, "what you're saying is that – hypothetically – if the demand is greater to have the coverage coordinated by the agent, then we forego any additional investment in the self-service model – or at least minimize it – in favor of providing a more robust solution for the agents?" Ann replied, "Absolutely – that's exactly what I'm saying." Oscar smiled again. He now had reached a new level of understanding about how all of this was going to help him solve SCI's financial issues, while at the same time delivering more value to the customer.

Short-Cycle Incremental Delivery

Two weeks after the initial meeting with the Bakers, everything was in place to launch. The larger group had gone through the now-revised MVP and created a "roadmap"—something that would serve as guidance to help them manage all of the moving parts. Ann had thought about what needed to be added to the website to begin to capture customer preferences and the leading indicators of progress and success along the journey.

What was left to do now was to think about how best to organize the various sets of expertise in the larger group, so things could start being worked on in a way that leveraged not only everyone's unique talents, but also the natural sequence of how the various pieces needed to come together. Everyone had agreed up until this point that they would develop the offering in a way in which pieces could be delivered out to customers; then they would take time to get feedback and learn, and then decide what the next piece of the offering would need to get developed. But now came the hard part of truly letting go of titles, organizational units, and the other trappings that had prevented the right amount of flow to occur to enable what they all envisioned to come to fruition.

It was a series of tough conversations, with highs and lows and much gnashing of teeth, but eventually the group decided on a structure of three smaller teams within the larger group.

These three teams were divided up into the following: 1) marketing & analytics, which focused on the changes that needed to be made to the existing website and analytic capabilities in order to test ideas and learn from existing and potential customers; 2) product development, which covered all of the work to actually define and create the product, including legal, underwriting, accounting, and the core systems processing teams; and 3) customer interaction, which provided the points of interaction for the client, whether through the online self-service facilities or via the agents.

The three teams kicked off their first work cycle (which Susan referred to as an "iteration") all at the same time. Each team had a "backlog"—a list of things to do—that they felt was both structured according to the scope of what the team was working on and sized to get done during the amount of time they had set for the length of an iteration, which was three weeks. Each of the teams attacked their backlog with unbounded energy. Each team also made some good progress in starting to finish some of their backlog items. But they soon realized that all was not as well as it seemed.

At the end of the first iteration, the marketing & analytics team had made the changes to the website that they had signed up for, but then realized that they needed to provide some time for customers to actually react in order to get feedback to help guide the development of the offering. This was discussed by the whole "team of teams," since it affected everyone, and a quick solution was devised to try and increase response rate. Howard suggested that they send out e-mails to all the customers alerting them to the new information on the website. This was promptly agreed to, and the e-mails were sent in a matter of days.

Another glitch that they ran into was encountered by team 3, the customer interaction team. They started off well, focusing on some of the easy user interface work that needed to get done for several of the features in their backlog. But they soon realized that, since a lot of the work around defining the offering in the internal systems had yet to be done, they had no way to test the connections they were programming from the user interface. This was a huge issue.

The group identified literally dozens of other "impediments" that were turning out to be serious obstacles for getting the initial MVP built.

Continuous Improvement

Each team was following the practice of doing a retrospective—a sort of review at the end of each iteration—to help them identify what they could be doing better. They realized, however, that it wasn't enough to simply evaluate each team individually. Now that they had multiple teams working on the same target, not only did they have individual team issues to manage, but also issues that spanned the teams and prevented the whole from working well.

The larger team came together one afternoon at the end of the next iteration to examine what was happening across the teams. Susan, the most senior scrum master in the group, opened up the session: "Well, we've been making great progress, everyone. But we also have our share of problems cropping up." Susan continued, "It seems one of the considerations we did not account for was the coordination that we need to have across the teams. Does anyone have any thoughts as to why that's happening?" Susan knew that a good scrum master always asks questions, even if they think they know the answer.

"Well," began John, "we are having all sorts of issues trying to sync up across the teams. I think it's because we plan the work well for our individual team, but then we don't share what that looks like across the larger group, so we get out of sync." Susan replied: "That's a really good observation; do you have a recommendation?" John answered, "Well, I think – after we spend some amount of time doing our team plans – if we could spend some time coordinating plans across the three teams, that would help a lot."

Rick, one of the developers on the product development team, spoke up: "I agree with John, but I also think we need to have a way to coordinate during the iterations. Look, we are constantly getting new information from the folks doing the product design based on the actuarial models that they are generating. Sometimes that creates changes in what we are going to deliver for the iteration that we are currently working on."

George, the product owner with the most experience, spoke up: "It sounds like what we need is a way to continuously manage what we will focus on by both planning ahead of, and coordinating during, each iteration." He continued, "My suggestion is that myself, Kira, and James (the other two product owners) work on an approach toward this continuous synchronization of what's being done across the three teams, so this way we maximize our accomplishments each iteration." The group agreed to giving it a try.

The group put the new activities in place for the next iteration. Now that the group was managing the work and impediments both within and across teams, things started to go much more smoothly. The synchronization before and during each of the iterations at the group level was helping build a better plan as well as do a much more comprehensive review of progress and problems as they occurred, so that plans could be adjusted during each iteration.

It wasn't long before the group was able to actually complete all of the work that had been defined as part of the initial MVP!

Learn from Linda—Shine the Light

Once the executives acknowledged that their department-centered operating model was a part of the company problem, Linda set out to show how implementing agile ways of working in the business can solve it. She set up half-day agile training workshops to educate the executives from delivery, marketing, finance, and channels about agile ways of working. During these workshops, agile experts taught the agile concepts in a highly interactive presentation with all agile jargon and paraphernalia. This sounds like formula for success, right? That's what Linda thought at first. That's what she hoped. Unfortunately, while these training workshops left her feeling good, the response from executives was surprisingly ambiguous.

"Thanks for the nice teaching session about all the new agile concepts," they said. "We understand that these are interesting theories that have been implemented in delivery but it's just not ready to work at the business level."

This subdued mood laced with skepticism was not the anticipated response.

Linda sat at her desk and stared out the window in disbelief. She had been geared up for a grand presentation by her team of agile experts just hours before; now she felt a cloud was hanging over the whole initiative. She kept asking herself what could have been done differently to prevent this pushback from the business after things had seemed so promising. Though these executives had so recently agreed they need to change their operating model to deal with the company problem, they remained unconvinced that agile ways of working were the answer.

With this resistance, Linda's path ahead was looking painful. Would the company remain mired in the old way of operating and miss the looming delivery deadline on this new initiative?

Linda reflected on the bad news.

"We had a great agile training session and now they are expressing major doubts about whether it can work across all their business areas," Linda thought.

"Ed Boderman in channels, Ann in marketing, all of them," Linda seethed in her mind. "For the past few weeks, I've been investing time to educate these executives, but they are having a hard time believing it can actually work as we are about to go big. None of it seems to make any sense."

So what was the issue? Although Linda had trained the executives in agile concepts, what she needed to do was to actually show the concepts working and driving tangible successes.

After thinking it over, Linda realized, "Instead of focusing on interesting training workshops about agile ways of working, I should switch to focus on how this will benefit each executive within his or her respective department. I need to speak only in terms of delivering customer value and then show them how success looks with a pilot demonstration. Instead of teaching agile concepts, I need show the executives how it helps accomplish their goals by delivering customer value while also solving the company problem."

Linda revised her approach to direct attention toward delivering customer value as the best solution to deal with the company problem. Linda later said, "I had to put the concept of customer value in bright lights so that executives could clearly see how to solve the company problem while also benefiting their department."

This would be a new way of thinking for the executives, because their measure of success has always been based on their department metrics, not on how much the customer values the company's products. Therefore, Linda articulated the benefits of an improved customer-oriented operating model and demonstrated them in a pilot. Instead of rolling out agile training classes for executives to learn agile ways of working, Linda got the executives motivated by shining the light on how each executive can benefit from delivering customer value with this new operating model.

CHAPTER

9

Agile Business Realization at Solar Corona

Absorbing the Results: Setting the Stage for Sharing

Six months had now passed since the pilot began. The MVP had been delivered two months prior, an astonishing achievement given SCI's previous track record. Davis requested a special review—broadcast across the company, as well as hosted in-person in the main office for whomever could attend there—to discuss the pilot: the results to date and the future of the new operating model as well as the pilot itself. Linda went to the team and asked how they wanted to present to Davis.

There were several suggestions, and after a bit of discussion, the group settled on an approach where every member of the team could volunteer to present a specific topic to Davis. This way, they could select something that they felt passionate about, and it provided them an opportunity to actually present to the company CEO!

J. Orvos, *Achieving Business Agility*, https://doi.org/10.1007/978-1-4842-3855-4_9

Linda went to Davis with the request to present the findings the way the team had asked to, and he didn't hesitate one bit, immediately saying "Of course!" Linda knew that this was just another example of why they were successful so far: it was the leadership provided by Davis that enabled everyone to take the risks that they did, including Linda.

Everyone started gathering in the auditorium on the main floor of the home office building. It was one of the few rooms that could hold all of the pilot participants, while providing the necessary equipment to broadcast the presentations. They had set the stage up with tables next to the podium, with all of the presenters seated. Once everyone had taken a seat, Davis went up to the podium to kick off the session.

"Good morning, everyone," Davis began. "I'm certainly excited to see all of you here today, and even more excited to hear your story about what you've accomplished for our company. I'm not one for long speeches, so let's get started!" Davis looked over at Linda, and she walked up the stairs to the stage and went to the podium.

"Good morning, everyone," said Linda. "I'm going to provide some initial context about the pilot and then I'm going to turn it over to my colleagues and let them highlight the various challenges and learnings that we faced and absorbed as the pilot progressed."

Linda proceeded to give the audience the context around how the pilot came to be, including the story of the now-infamous budget meeting where Oscar sounded the alarm. She provided a colorful summary, but quickly covered all of the key events and decisions that had gotten the group to where it was, including the eventual rollout of the MVP. "But," Linda added, "things did not always go smoothly, and we struggled to find our footing. What I would like to do next is to have our panel members here onstage take you through some of the challenges and learnings that we experienced." Linda looked over to Ann and signaled for her to take it over.

Customers

"Good morning, everyone," said Ann. "All of the people you see here at the table onstage have volunteered to present certain aspects of the pilot to you today, not because we are experts, but rather because we have been profoundly changed by what we have experienced over the last six months and we want to share that with you. My colleague Ed and I come from marketing and channels, respectively, and when Linda initially approached us, we thought we knew what customers wanted."

"That's right, Ann," Ed added. "I thought I could make all the decisions about what our customers needed. Boy, was I wrong." "And that goes double for me, Ed," Ann exclaimed. "With my background in marketing, I thought I knew how to identify customer needs, using methods that have been around for years. What I think both of us have come to realize is that we couldn't have been more wrong."

Ann and Ed then told the story of how they both were vested in their own ways of working, but each had been transformed by the journey that Linda had taken them on. This included the story of how the group initially defined the MVP, only then to meet with the Bakers and receive totally new insights from them. "We thought we knew what they wanted, but we learned throughout the pilot that as you provide bits and pieces of a solution to someone, they get better at figuring out what's really important to them," said Ann. "The combination of working directly with our customers and being able to go through cycles where we deliver things they can actually start to use was transformative," Ann continued. Then Ed exclaimed: "I agree! I have had a light bulb go off that has changed the way I think about customers forever!"

Teams

Linda saw that Ed and Ann were finished, and so she asked Chuck to go next. "How's everyone today?," Chuck asked the audience. "I'm here to tell you about my own transformation that occurred during this pilot. For me, the most powerful aspect of this whole journey has been how the teams have come together to get the job done. Most of you know me – I've worked at this company for over 25 years and I'm an old-school kind of guy; pretty stuck in my ways. I came from the school of thought that you have to have a fairly defined management structure, and everyone needs to be told what to do. Boy, I couldn't have been more wrong."

Chuck continued, "The first thing I saw when we started seemed like chaos to me, I've got to be honest. But everyone kept telling me to be patient, so I held my tongue. And what I saw happen was truly amazing. The people on the teams – although initially struggling to find the right approach – eventually found it and started firing on all cylinders! Not only did they figure out what to do, but they also figured out how to work together as a team, and then eventually, how to even coordinate across teams – all without being told what to do. And probably the thing that impressed me the most was the fact that they held each other accountable for what needed to get done. I'm truly humbled and will never doubt the power of teams to self-organize and self-manage."

Learning Fast

Steve was next in line to talk about his experience on the pilot. "You know, what I learned goes against everything I knew about running infrastructure," Steve said in a tone of disbelief. "Running the infrastructure organization, I was convinced that the only way was to plan, recheck your plan, and then plan again so that you knew precisely what you needed and then went and built or purchased it," Steve said. "Like the other panelists up here today, I'm here to tell you, I was forgetting something really important: you can really never be 100% certain of anything, and the amount of time you spend trying to get there may mean that you miss an opportunity to do something great for your customers."

Steve continued: "I saw the pilot challenge the way we provision infrastructure, as well as a host of other activities that I thought you traditionally needed to spend more time planning. What I learned is that you don't know what you don't know. Sounds simple when you say it, but it fundamentally changes the way you think about how you approach getting stuff done. Instead of trying to plan everything ahead of time, you lean more towards taking action, watching what happens, and then adjusting and moving forward. If customers don't know what they want until they start to see it, then we need to let them learn fast – while we learn right alongside them."

Measurement

Wrapping up the set of presentations was Oscar. "Being the CFO, my life is about numbers," said Oscar. "What I learned throughout this pilot is that I haven't been looking at all the numbers that I should have been," Oscar mused. "I've spent my days focused on things like ROI, ROA, EBITA, and dozens more measures that I was taught about in school and then asked to produce throughout my career. But there's more to measurement than – and I learned a new term here – *lagging* indicators; just as important, and at times even more so, are *leading* indicators."

"Lagging indicators tell us where we've been," Oscar continued. "With the pilot, we've embarked on an entirely new journey; one where we need to have instruments that tell us where we are headed and how well we're doing on our journey *before* we get there. That's what the concept of leading indicators covers – looking at trends that help provide direction and help us steer our company down the right path, and to make adjustments before it's too late."

After the presentations were all said and done, Davis asked Linda to come up to his office for a chat. "Linda, I can't tell you how impressed I am with the results of this pilot," Davis said. "And by results, I'm not focused on the specifics around the transportation insurance offering that you delivered, as much as I am impressed with the way you pulled the organization together to all focus on a common goal."

"Well, thank you, Davis," Linda said sheepishly. Linda never was one to look for credit for anything that she worked on. "It truly was a team effort," Linda added.

Smiling, Davis looked at her and said, "Yes – but they had a great leader. I'm very proud of what you've done here. I think it's time you and I start discussing your future at SCI; I think we're going to need someone to drive our new operating model adoption!"

Linda smiled, and the two leaders shook hands.

Learn from Linda—Beyond the Pilot

Upon concluding the successful pilot, Linda was confident in her abilities to sustain and expand newly minted agile business practices throughout the organization. After all, the team that performed the pilot had made tremendous progress. In the past, it was anyone's guess whether software would work and deliver value to the customer. But now, the agile team delivered working software that provided real customer value in response to what customers actually desired. Naturally, the pilot team was excited about their success and their enthusiasm began to spread to other teams.

Linda believed that this enthusiasm would help drive expansion of this new way of working throughout the organization. Yet her company executives were not nearly so confident. They shared her enthusiasm, but did not fully believe that they could expand this pilot to a larger scale in the company. Linda asked early and often for commitment; all she earned in return was elevated stress levels. Linda was badgering the executives with her repeated, though polite, requests to secure their commitment, when what the business executives really needed was a structured approach to expand this pilot success.

Even though they were excited about the success of the experimental pilot with the single team, the business executives' focus was on the much bigger picture. And they couldn't really imagine how to expand that pilot success to serve the full company. Meanwhile, by assuming the small, experimental pilot would naturally expand itself into the larger organization, Linda inadvertently came across as cavalier and naïve. She lost some of her credibility. Linda concluded that this presumptive approach will fail in gaining long-term, organization-wide commitment for an agile business. The last thing she wanted to do was pressure the executives and risk blowing all that hard work that produced the successful agile pilot.

By creating a realization plan to reinforce the new agile operating model practices on a larger scale, Linda found success in overcoming executive concerns. As Linda explained, "Even after we had executive excitement from the small successful pilot, we still did not get a commitment to do this across the company. I was able to overcome this by creating a comprehensive reinforcement plan that gave them confidence."

CHAPTER

10

Learning from Solar Corona

After relaying Linda's story to a couple of agile transformation consultants, I asked, "So how'd she do it?"

"Simple," said Michael, a lead consultant. "She had support from business executive leadership."

Dan, another respected agile consultant, chimed in, "Oh, yeah, I always tell my clients that if they don't get executive buy-in they cannot have positive change in their organization."

I pressed them for more specifics, especially since I knew Linda's success had happened in an industry-leading company with executives set in their ways. "But how did Linda actually earn this executive buy-in?," I asked.

"Well, she had company executives who trusted her and who were willing to change," said Dan.

"So how did Linda build that trust?," I continued. "Did she just ask around, hoping to find someone who wanted to work with her? Was it basically luck?"

Silence. Blank stares. Then a vague, slightly defensive reply from Michael: "Yeah, well, she found a courageous individual willing to take a chance and support her."

The conversation ended there, and I walked away thinking, "I guess Linda is an outlier – the purple unicorn who made an agile business happen." Sadly, none of the agile experts really knew the secret to how it had happened.

J. Orvos, *Achieving Business Agility*, https://doi.org/10.1007/978-1-4842-3855-4_10

None of it seemed to make any sense. I replayed Linda's story in my head over and over again, determined to learn how she succeeded. I dedicated time to conducting interviews with Linda as well as numerous other agile experts and thought leaders. During my research, I continued to hear the same wisdom from my agile consulting expert friends: that the key to adopting agile ways of working with the business is getting executive buy-in to make the change. But to my dismay, there did not seem to be anyone offering a real strategy to earn that.

Then I realized something. Linda had not actually followed a systematic approach rooted in her agile expertise. Her success story was about the unique ways she communicated and handled her interactions with business leaders to earn their trust. This was the sweet spot and yet a gray area as well. Someone needed to capture that story and replicate it. So, I continued my inquiries and started writing—and the basis for this book emerged.

So, Linda's story ends well. By changing her approach, Linda was able to skyrocket from respected executive to a rising star. Better yet, while she was indeed gaining attention, she was also more fulfilled and having fun along the way because this change shifted her focus away from pushing an agenda and onto helping her executives colleagues solve problems.

I have to thank Linda for taking the leadership role to work through her difficult challenge. At the time, it seemed the only rhyme or reason for her successes was a bit of luck and determination to have others buy into her goal to adopt an agile business. Like so many professionals, she thought she was just fortunate for making this happen. With the future of an agile business at stake, though, I felt compelled to take a long, hard look at the underlying formula of success in this case. What I discovered then forced me to shift my way of thinking and realign the approach to an agile business. This process was borne out of that journey.

Meet People Where They Are

As you have read throughout Linda's journey, she continuously changed her communication style to be aligned with the executives' mindsets. Linda had to think about how her message was perceived based on a given executive's mindset, rather than what she herself wanted. At times, she realized, she had been overzealous and presumptuous to present how great agile ways of working were to anyone that would listen. By prematurely assuming that her audience was interested in agile, Linda had gone barreling in to deliver her pitch. She started talking extensively about the wonders of agile ways of working and success in delivery, aiming to demonstrate her expertise on the subject, but was instead raising the stress levels of her audience. Moreover, badgering her executives with polite, repeated prompts to change their process had inadvertently undermined their trust in Linda. As a result, Linda had lost her connection with the executives and her efforts were failing.

Linda's mistake of "jumping ahead" is a common problem for many agile transformation advocates striving to motivate business executives to change. If you move too fast and rush ahead, "guns blazing," to adopt agility in the business, you will create resistance and sometimes resentment. Understandably, business executives are not all that interested in learning a new set of agile languages and practices until they realize and recognize its worth, and how it will help them in achieving their own goals. Moreover, jumping ahead puts the executive in the uncomfortable position of having to translate what you are trying to say. That's because the message is not in the business language that they are accustomed to. Everyone needs a personal motivation before they will take time to learn the details of a new process.

Learn from Linda

Once Linda realized that her communication style needed to align with how business executives think, she made a serious change. She flipped her approach and started talking to the executives in their own business terms, based on their particular individual mindsets. The result was a breakthrough. From that point forward, Linda could establish trusting relationships and effectively usher the executives toward achieving business agility.

All of these steps Linda took in these strategies are simply conversations that align with the way business executives think and make decisions. You have to earn their buy-in and trust based on how your conversations align with their own business goals and perspectives. When you focus on conversations with the executive that stay aligned with their thinking, you can bring them elegantly through their decision process toward change. By using the strategies from this book as a guideline for your conversations, you will show them how they can benefit, which will motivate them to join your movement to change.

PART

3

The Takeaway - Defining Your Company's Journey

CHAPTER

11

Importance of Alignment

To summarize, the four actionable strategies to achieve business agility are as follows:

"**Sound the Alarm**" provokes a response through a statement that calls attention to a particular company problem the executives may have been missing.

"**Look in the Mirror**" educates and collaborates with executives to reveal how their operating model is contributing to the overall company problem and how they will need to be a part of the solution.

"**Shine the Light**" creates the vision for change by highlighting what it looks like to deliver customer value with a pilot that demonstrates success from a customer's perspective.

"**Agile Business Realization**" establishes a reinforcement model to hold executives accountable to change and make it stick over time.

Align with Executive Thinking

Now that you know these four strategies, utilize the right one at the right time to align with the way business executives think and make decisions (Figure 11-1). When you focus on conversations with the executive that stay aligned with their thinking, you can elegantly guide them through their

J. Orvos, *Achieving Business Agility*, https://doi.org/10.1007/978-1-4842-3855-4_11

decision process to effect change. By using the strategies from this book as a guideline for your conversations, you will show them how they can benefit, thus motivating them to join your movement to change.

Align With Executive Thinking

EXECUTIVE MINDSET	**Awareness of any problem?** "I'm doing just fine and can meet my objectives."	**Knowledge about the problem?** "I don't know how this problem applies to me."	**Ability to solve the problem?** "I need to see how I benefit from making a change."	**Company success plan?** "How will we achieve business results?"
YOUR STRATEGY	**Sound the alarm** "Our company is facing new challenges."	**Look in the mirror** "Your process is a part of the problem."	**Shine the light** "This is how success looks."	**Agile business realization** "Here is the reinforcement plan."

Figure 11-1. Align with executive thinking

Don't Go Too Fast

At times, you may be overzealous in teaching the mechanics of agile ways of working. Avoid this temptation; don't fall into this trap! By assuming an executive is aware of a problem, and seeking to learn more about how to improve their company/department operating model, don't immediately proceed to share the wonders of agile ways of working and success in their delivery department or other organizations, aiming to demonstrate your expertise on the subject.

Avoid the mistake of "jumping ahead," which is a common problem for many agile transformation advocates striving to motivate executives to change. This will backfire and raise executives' stress levels. If you move too fast and rush ahead "guns blazing" to adopt agile in the business, you will create resistance and sometimes resentment. Moreover, badgering your executives to change their company/department operating model, before they have the awareness or knowledge of the problem, will inadvertently undermine their trust. Understandably, business executives are not really that interested in learning a new set of agile languages and practices until they realize and recognize it's worth it for them to improve their own goals. Moreover, jumping ahead puts the executive in the uncomfortable position of having to translate what you are trying to say. That's because the message is not in the business language to which they are accustomed. Any person needs to have a personal motivation before they take time to learn the details of a new process. The result is you risk losing your connection with the executives and your efforts will fail.

Slow and Steady

On the other hand, there's also a risk of being too cautious and assuming executives don't realize there is a problem in the first place. More than likely, the executive already knows about their own business, so you will not be telling them anything new, as they likely have an "awareness" of their own business problems. Tread lightly here since they will wonder why you are trying to teach them something they already know about, namely, their own business pressures. What they might lack is an understanding and knowledge that they are part of the problem. If you spend excessive time explaining a problem the executive already knows about, you will create frustration. At times, you may misunderstand the mindsets of the executives and fail to give them credit for what they already know. You might dwell too long on the problems the company is facing and miss the points that matter most to the executives, failing to offer insights and knowledge into their process constraints or action steps they can take to solve these problems. The already impatient business executives will interpret your lack of this understanding as a sign of inexperience, which leaves them feeling your meetings are a waste of time.

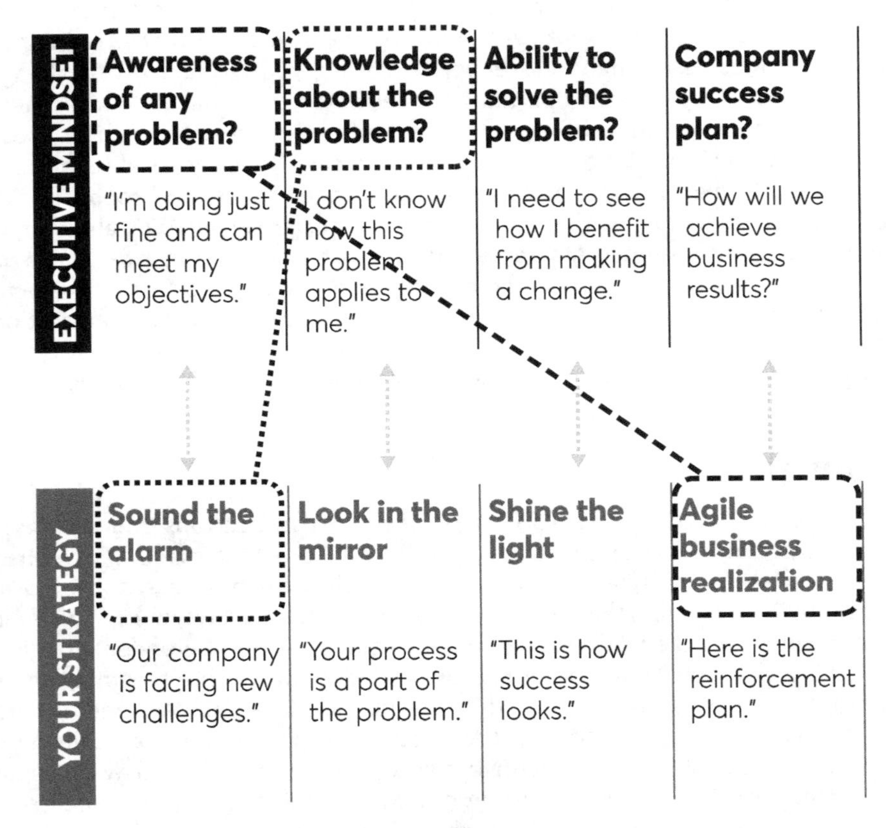

Figure 11-2. Don't go too fast or too slow

Selecting Your Strategy: Mindful Surveillance

Deciding on the right communication strategy is based on the mindset of each executive. To understand the executives' mindsets in your company, ask the following three questions to reveal their awareness and knowledge about the company problem and put together the executive mindset map (Figure 11-3).

Executive Mindset

Is there **Awareness** that a company problem exists?

Is there **Knowledge** that they are a part of the company's problem?

Is there an **Ability to Take Action** and solve the company's problem?

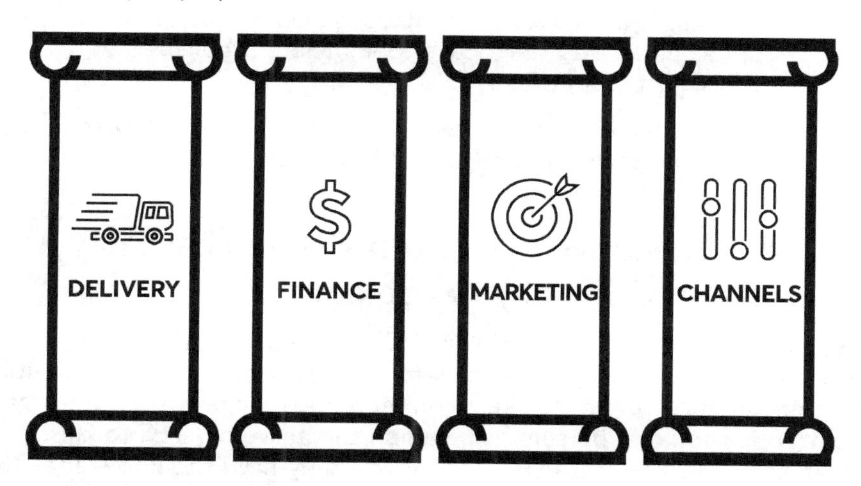

Figure 11-3. Executive mindset

(Since executive mindset varies from person to person and could change over time, you will need to keep track. It's important that you understand and continually survey the different attitudes and thoughts of your executives in the four pillars.)

Is the Executive Aware of Your Company's Problem?

Start by asking your first question that measures the executives' awareness about the company problem. From his or her answer, you will learn whether or not the executives know there is a problem dealing with the new customer and competitive pressures on the company. Ask, "Do you understand the pressures our organization is facing as we deal with disruptive competitive pressures? Do you agree there is a company problem?" His or her answers to these questions will determine our communication strategy.

If they answer NO, give them "Awareness" that a problem exists: Sound the Alarm (see Chapter 1).

- If the executives do not have the awareness of these company pressures and feel they are doing fine because they are meeting their department objectives, sound the alarm and help them recognize that the company is vulnerable to being crushed by more agile competition. Through your lens, you now have this treasure to share. You will be doing them a favor for contacting them so they can be proactive, instead of reactive to imminent problems that will impact them ahead.

If they answer YES and they are aware that a company problem exists, ask the next question about their knowledge about the problem.

Does the Executive Have Knowledge That They Are a Part of the Company's Problem?

Most of the time, executives answer YES to the awareness question and know about the company pressure with emerging disruptive competition. The company pressures are likely common knowledge within the organization, so it shouldn't be surprising when the answer is YES to this question. In this case, ask the next question: do they have the knowledge that this applies problem to them? Ask, "Do you agree your department needs to change to deal with that problem?" If they answer NO, give them the "Knowledge" of how they are a part of the problem: Look in the Mirror (see Chapter 2).

- Although executives may be aware a problem exists with emerging company pressures, they may not know how this applies to them. As a result, they likely have initiatives underway that ignores the problem using their old, rigid company/department operating model. They think they have nothing to do with the company problem. In this case, they are aware of the company problem of increased disruptor competition, but do not have the knowledge that they must be a part of the solution to help fix it. In this situation, your communication strategy is to have the executives Look in the Mirror (see Chapter 2) and face the fact that their company/department operating model is a part of the problem.

If they answer YES, and they have the knowledge that they are a part of the solution to the company problem, ask the next question about their ability to take action to solve the problem. Ask, "Do you understand how your department will enjoy benefits under a new customer-centered operating model?" Give them the "Ability to Take Action" and conduct a pilot that highlights the benefits of a customer-centered operating model: Shine the Light.

Does the Executive Have the Ability to Take Action and Solve the Company's Problem?

In this situation, your executives are aware of the company problem and also have the knowledge that the company problem applies to them. They need your guidance to learn how to take action. In this case, Shine the Light (see Chapter 3) to show them how to change by conducting a pilot of the customer-centered operating model in action.

Engage Trouble Spots

Be aware of any troubles spots you encounter. Your surveillance may reveal an executive opposed to your efforts—this is your hidden opponent. You may never have met them during your previous interactions or they may have attended your meetings and conference calls but remained silent. These opponents can be against change for any number of reasons. They may see your proposed change as a distraction to their normal ways of doing things. They may believe that stability outweighs improvement efforts and wish to keep the status quo. Regardless, the opponent stakeholder, predisposed against your proposed customer-centered operating model, may choose to not engage or participate.

To determine how the executive mindset compares to others, ask, "What will your staff and colleagues be saying about your views? Will they all agree? Will they be comfortable with your perspectives? Is there anything I should know that might prevent us from moving forward with our discussions?" These questions may seem awkward but vital to avoiding any surprise setbacks.

If your mindset surveillance uncovers resistance to change, set a meeting with them one on one when they can express their concerns and you can address them directly. Begin the meeting by respectfully asking the opponent, "I value your thoughts on this. Can I better understand if you see risks stemming from our conversations?" Then use your strategies as a tool to answer these concerns.

The answer to this question will reveal if they don't have the awareness, knowledge, or ability to take action regarding the company problem. Based on their answer, simply go back to that particular communication strategy and implement this with the opponent one on one. Your opponent has been respectfully engaged and the objection can be neutralized.

Let's Review

- Before engaging with executives, understand their current mindsets to determine your strategy
- Align your strategy with the mindset of each executive with whom you engage
- If you misalign and push too fast or go too slow, you may lose support from the executive

CHAPTER

12

Your Journey Ahead

This book rests on the premise that you can create change in your organization and achieve business agility. By implementing the four strategies in this book in alignment with the mindset of your executives, you can succeed. Now that you have read and understand the four strategies of enabling business agility you are ready to create real and impactful change in your company. By enabling all four pillars to pivot together, they will achieve agility which will help your organization defend against emerging disruptive competitors in a digital world. It will also gain you personal visibility that may propel you to the next level in your career.

Of course, adopting agile is all about leading change, but as stated throughout the book, creating change is about much more than forming trusting relationships. It's about helping other people connect with a new ideal state so they want to improve their situation. It's about helping them realize what the problems are and solving those problems.

As you go about implementing the approaches in this book, bear in mind this represents a new and exciting way of doing business in your organization. With information and knowledge comes power, and with power comes influence. The ability to identify and harness the power of agile will translate into increased sales and more profit for your company. You are now on your way to a more rewarding career.

J. Orvos, *Achieving Business Agility*, https://doi.org/10.1007/978-1-4842-3855-4_12

As you aspire to be an agent of change, your colleagues will judge how you interact with them. More than ever, creating trusting relationships with others is critical to your success. There's no way around this: executive decisions are influenced by their conversations and interactions with you. You've now learned the process of creating an agile business equipped to pivot and be able to truly sense and respond to change. You have attained the mission of elevating business agility to business as usual.

APPENDIX

A

Quick Reference Conversation Guide

Your impact statement will sound the alarm to executives that competitive disruption and customer needs are changing the business landscape, and that to remain competitive, your company must change with it. Your statement gives you the chance to identify the imperative changes and actions the company must make and the business consequences of *not* taking action.

J. Orvos, *Achieving Business Agility*, https://doi.org/10.1007/978-1-4842-3855-4

If you are in the **Delivery** organization, you will need to emphasize that success is dependent on building the *right* products that customers need (Figure A-1).

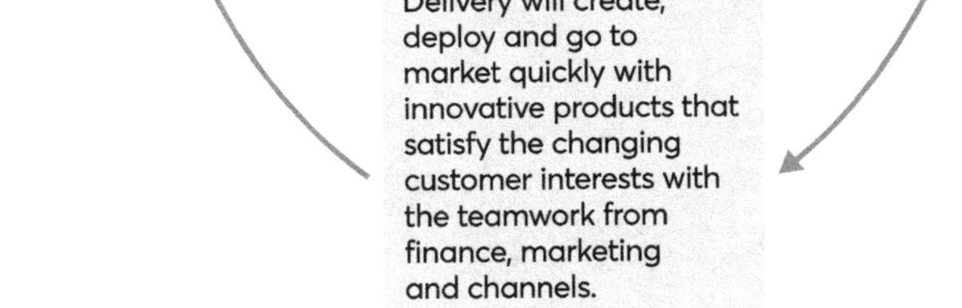

Figure A-1. Example impact statement for Delivery

If you are in **Finance**, you must stress profit margins and funding smaller and more frequent work items that are designed to changes in the market (Figure A-2).

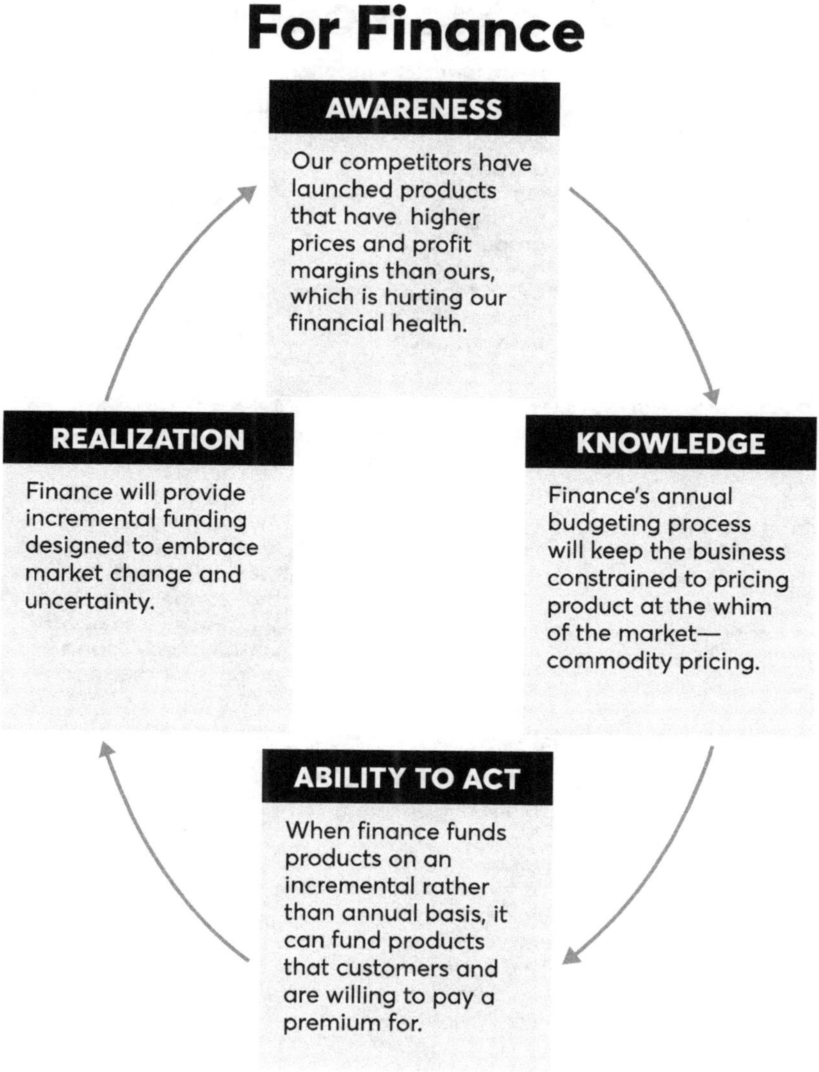

For Finance

AWARENESS

Our competitors have launched products that have higher prices and profit margins than ours, which is hurting our financial health.

KNOWLEDGE

Finance's annual budgeting process will keep the business constrained to pricing product at the whim of the market—commodity pricing.

ABILITY TO ACT

When finance funds products on an incremental rather than annual basis, it can fund products that customers and are willing to pay a premium for.

REALIZATION

Finance will provide incremental funding designed to embrace market change and uncertainty.

Figure A-2. Example impact statement for Finance

If you are in **Marketing**, the emphasis is on identifying market demands, being able to pivot quickly on them, and then communicating them to the rest of the organization (Figure A-3).

For Marketing

AWARENESS

Our competitors are attracting new customers that we are missing because our products (and messaging) are steps behind the market. Our products are not as relevant as our competitors'.

KNOWLEDGE

The difficulty arises when the company's marketing messages are constantly a step behind those of competitors—in the crowded waters of this bloody "red ocean," with so many sharks going after the same customer.

ABILITY TO ACT

Marketing will promote the right products and features (internally and externally) and develop relevant messaging to customers and improve the quality of potential leads to increase the sales conversion rate.

REALIZATION

Marketing will engage with delivery, finance and channels to adjust customer messaging based on feedback from the market, in favor of pivoting rather than sticking to a long-term irrelevant plan.

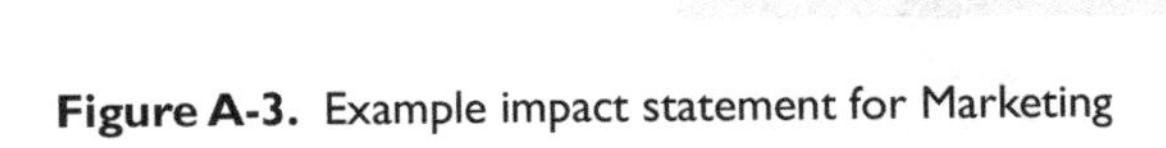

Figure A-3. Example impact statement for Marketing

If you are in **Channels**, you must emphasize the voice of the customer; your new role is to champion products and features that your customers and the market demand (Figure A-4).

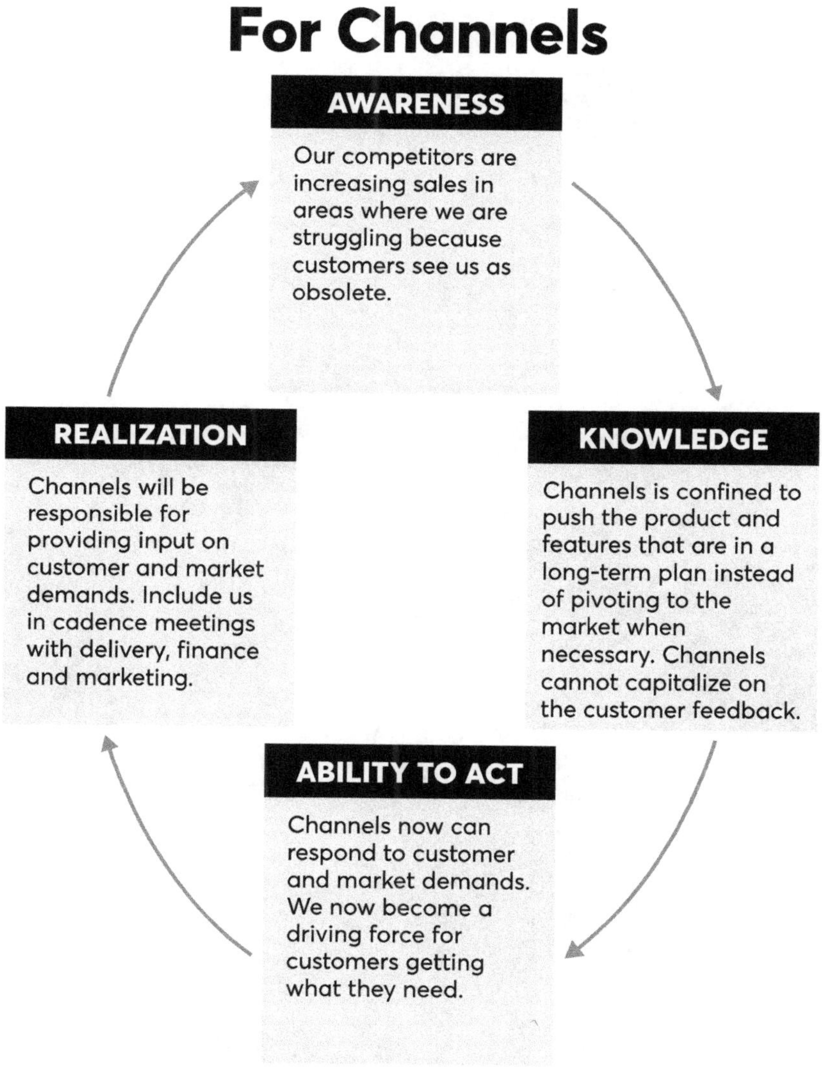

Figure A-4. Example impact statement for Channels

I

Index

J. Orvos, *Achieving Business Agility*, https://doi.org/10.1007/978-1-4842-3855-4

C

P

Q

R

S

T, U

V

W, X, Y, Z

CPSIA information can be obtained
at www.ICGtesting.com
Printed in the USA
BVHW04s0659231018
530984BV00004B/14/P